Research and Design of Hydraulic Impact Mechanisms

Qinghua He

Research and Design of Hydraulic Impact Mechanisms

Springer

Qinghua He
Sunward Intelligent Equipment Co., Ltd.
Changsha, Hunan, China

ISBN 978-981-96-4863-4 ISBN 978-981-96-4864-1 (eBook)
https://doi.org/10.1007/978-981-96-4864-1

This Springer imprint is published by the registered company Springer Nature Singapore Pte Ltd.
The registered company address is: 152 Beach Road, #21-01/04 Gateway East, Singapore 189721, Singapore

Preface

Hydraulic impact mechanism is the key component of the new generation of rock drilling equipment—hydraulic rock drills and hydraulic crushers, with significant characteristics such as high efficiency and energy saving, widely used in mining and civil engineering.

All moving parts of the hydraulic impact mechanism are always in a severe variable speed motion with acceleration (deceleration) up to tens or even hundreds of times that of gravity acceleration during operation. The control valve should switch the oil circuit of large opening within 1–2 ms. Therefore, the moving parts will generate a large inertial force. The working oil pressure of hydraulic impact mechanism mainly depends on the inertia force of the moving parts, and the pressure and flow rate vary greatly. These characteristics pose certain difficulties for the research, design, and manufacturing of hydraulic impact mechanisms.

This book is written by the author with the summary of years of research results, further theorization, systematization, standardization, and application. There are seven chapters, mainly including: Part I, a linear model of hydraulic impact mechanism is built, and the "Three-Step Analysis Method" is proposed; Besides, two dimensionless variables are introduced, β (the ratio of the acceleration during the return acceleration phase to that during the impact stroke phase) and β_1 (the ratio of the acceleration during the impact stroke phase to that during the return deceleration phase); With these, various parameters and calculation formulas are derived; Part II, an in-depth analysis was conducted on the factors affecting energy loss and efficiency, the relationship between flow rate and pressure, the charging and discharging oil volume of high-pressure accumulator, and the reasonable selection range of β, illustrating the motion law of hydraulic impact mechanism more thoroughly; Part III, taking into account factors such as oil compressibility and hose expansion, a nonlinear mathematical model and simulation model of the impact mechanism were established, a simulation program was developed, and a quasi-constant acceleration numerical calculation method and correction method were proposed, effectively improving the accuracy of simulation calculations; Part IV, the function and calculation of the return oil accumulator is analyzed, and the causes and prevention measures of pressure fluctuations and cavitation in hydraulic impact mechanisms are explored.

Since the first publication in 1995, the original book has been used as a graduate textbook for related majors and has received unanimous praise from peer experts and scholars. Hundreds of scholars have cited it in their compilations. In the revised edition published in 2009, the system of the original book is retained, the original content is checked, and outdated contents are removed. In the English version, in addition to the deletion of outdated contents, the simulation program in the book is also rewrote with Python, which obtain simulation curves easier and more convenient analysis. Meanwhile, three computer programs are attached in this book, including "calculation program of hydraulic impact mechanisms by three-step analysis", "simulation program for hydraulic impact mechanisms", and "simulation program for hydraulic impact mechanisms without high-pressure accumulators".

Associate Professor Yong Guo and Prof. Jianxin Zhu, made significant contributions to the tests in this book. Associate Professor Yanling Gong has completed the conversion and proofreading of the simulation program from the original BASIC language to Python language during the republishing of this book.

Although this book has been repeatedly reviewed, there are inevitably shortcomings and errors. We sincerely welcome any criticism and correction.

Changsha, China
December 2024

Qinghua He

Contents

Chapter 1
Introduction

1.1 Overview

Impact is a common physical phenomenon. For an impact to occur, two conditions must be met: first, there must be an impacting body with a certain amount of kinetic energy (velocity), and second, there must be a target that can rapidly decelerate the impacting body. Upon careful analysis of the impacting bodies, it can be observed that: some impacts are disposable, such as bullets; some impacts, though used multiple times, have irregular and intermittent motion, like a hammer; while others have continuous and regular motion.

Utilizing the impact force to acquire prey and create tools can be considered as one of the earliest productive activities of human beings. As the progress of human civilization, the earliest manual stone-based impacts have long been replaced by a wide variety and complex impact machinery. This book focuses on a type of impact machinery related to the fragmentation of brittle objects such as rocks or ores.

In mining, hydropower, railway, and other production and construction activities, the amount of work involved in separating rocks (or ores) from their associated matrix (ore deposits, rock formations, etc.) is huge. Despite the various methods available for separation, the dominant method remains drilling and blasting on the rock (or ores) surface, commonly known as the drilling and blasting method. Currently, there are many methods and corresponding equipment for drilling blast holes, among which mechanical drilling methods can be classified into three types: impact-based, rotation-based, and combination of rotation and impact. The impact-based drilling method is the earliest and most widely used to this day.

In addition, the process of using impact to break larger brittle objects such as rocks (or ores) and concrete into smaller pieces, known as secondary crushing, is also a production operation with a considerable amount of work involved.

After centuries of using manual hammers and chisels, the pneumatic rock drill was invented by the Swedes in the nineteenth century during the excavation of the

Q. He, *Research and Design of Hydraulic Impact Mechanisms*,
https://doi.org/10.1007/978-981-96-4864-1_1

Mont Blanc Tunnel. This invention brought about a historic revolution in the excavation of blast holes, and it can be seen as a significant advancement in impact tools. Over the next 100 years, pneumatic rock drills and other pneumatic impact tools greatly improved in terms of variety and performance (Guo 1993; He et al. 1990; Xia 1985, 1990). Through extensive research and application, the theoretical study, structural design, and manufacturing processes of modern pneumatic impact mechanisms have reached a state of perfection. With the advancement of hydraulic technology, hydraulic machinery that uses high-pressure fluids as a transmission medium has become more widely utilized. In comparison, compressed gas as a transmission medium has much lower energy density and significantly lower energy utilization efficiency (Xia 1985). It is evident that pneumatic impact mechanisms not only face difficulties in significantly increasing impact energy but also have large energy consumption and high noise levels that exceed environmental standards. These weaknesses become more conspicuous in modern industrial production, which emphasizes high efficiency, low energy consumption, and environmental protection. For these reasons, the development of hydraulic impact mechanisms began. Since the 1970s, hydraulic impact mechanisms have made significant breakthroughs in practical applications (Xia 1985).

In 1970, France took the lead in developing the world's first practical hydraulic rock drill, which was quickly mass-produced and widely promoted (Guo 1990; Xia 1985). Subsequently, countries such as Sweden, Germany, the United States, Finland, and Japan developed various models of hydraulic rock drills or hydraulic breakers and introduced them to the market. Due to the high efficiency and energy-saving characteristics of such equipment, more than a dozen countries, along with dozens of manufacturers and research institutions worldwide, have actively engaged in research and development. As a result, hundreds of hydraulic rock drill and hydraulic breaker products have been introduced, with some undergoing several generations of updates (Cao et al. 1989; Xia 1985, 1993). After rounds of competition and elimination, the hydraulic rock drilling equipment produced by Sweden's Atlas Copco and Finland's Tamrock has become the most widely used, accounting for more than 65% of the global market share.

Hydraulic rock drills and hydraulic breakers, with hydraulic impact mechanisms as their main working components, have become a new generation of highly efficient and energy-saving rock drilling equipment. Compared with pneumatic rock drills and pneumatic breakers, hydraulic rock drills achieve drilling speeds 1 to 2 times faster, while the crushing capacity of hydraulic breakers has also doubled. Additionally, their energy utilization efficiency has increased from 15 to 40%–50%, and noise levels have been reduced by 10–15 dB, significantly improving productivity (Guo 1993; Wang et al. 1987; Xia 1985; Zhang 1991; Zhao 1996).

China started slightly later than other countries, beginning research on hydraulic impact mechanisms in the late 1970s. Through a combination of technology importation and independent development, many critical experimental and technical challenges were overcome, leading to rapid progress. Significant innovations were made in working mechanisms and design methodologies. To date, more than a dozen institutions have developed and manufactured dozens of models of hydraulic rock drills and hydraulic hammers, which have been successfully applied in energy development, urban construction, and tunnel engineering projects (Guo 1990; He et al. 1990; Jiang 1995; Li 1991; Xia 1987, 1990, 1993).

Based on comprehensive theoretical and experimental research, my organization, Central South University, has taken the lead in developing the first large impact power hydraulic rock drill in China. Besides, we have also successfully developed domestically manufactured underground fully hydraulic roadway drilling rigs and open-pit hydraulic drilling rigs. It can be said that we possess mature and unique technologies and experiences in the research and application of hydraulic rock drilling equipment.

1.2 Working Principle of Hydraulic Impact Mechanisms

Since the introduction of hydraulic impact mechanisms in rock drills by the French company Montabert in 1970, numerous new concepts and related patents regarding hydraulic impact mechanisms have been proposed both domestically and internationally. Although the forms of these concepts vary, their goals are the same, solving the problem of achieving reciprocating motion of the impact piston from the impact stroke to the return stroke. This involves utilizing high-pressure fluid (in some cases, high-pressure gas enclosed within the impact mechanism) to drive the impact piston and enable its reciprocating motion. During the impact stroke, the result is the delivery of impact energy to the target object, while the return stroke prepares for the next impact. Among these concepts, there is a clever design that utilizes the compressibility of hydraulic oil to create self-excited vibration. This type of hydraulic impact mechanism, which consists of only one moving body, namely the impact piston, is known as a valve less hydraulic impact mechanism. However, due to its large size, heavy weight, and low impact energy, it quickly disappeared from the market. Hydraulic impact mechanism which utilizes control valves to switch the flow paths for the impact piston between impact and return strokes takes the dominant place. With the exception of a few utilizing a so-called forced diversion scheme, where the directional shift of the control valve is achieved through external forces such as hydraulic pressure or electromagnetic force, the majority of systems adopt an automatic feedback method. The directional shift of the control valve relies on the movement of the impact piston to open and close the flow channels, which in turn switches the fluid flow path of the piston. This process repeats itself, and with the aid of position feedback, it automatically achieves periodic impact motion.

In hydraulic impact mechanisms, the pressure chamber that drives the piston during the return stroke is generally referred to as the 'front chamber' (near the impact end), while the pressure chamber that drives the piston during the impact stroke is called the 'rear chamber'. Existing hydraulic impact mechanisms includes three types based on the variation of pressure in the front and rear chambers.

One is the 'rear chamber control' impact mechanism, which has a front chamber that operates at constant pressure, and a rear chamber that serves as the control chamber with alternating high and low pressures. It is referred to as the 'rear chamber control type' or 'rear-control type' in this book, which is currently the most widely used in practical applications.

Another type is the 'dual chamber control' mechanism, where both the front and rear chambers alternate between high and low pressures. It is referred to as the 'dual chamber control type' or 'dual- control type' mechanism in this book and is the second most commonly used mechanism, accounting for a significant proportion of applications.

There is also a type of impact mechanism where the rear chamber operates at constant pressure, while the front chamber alternates between high and low pressures. However, this type is rarely used in practice.

The rear-control and dual-control types hydraulic impact mechanisms will be discussed in this book.

Whether it is the rear-control type or the dual-control type, the movement of the impact piston of hydraulic impact mechanisms is divided into two phases: the return stroke and the impact stroke. The return stroke can be further divided into two phases: return acceleration and return deceleration. The basic working principle of these two hydraulic impact mechanisms is introduced below, illustrating how the movement of the valve spool coordinates with the motion of the impact piston to achieve the reciprocating motion of the return stroke and the impact stroke.

1. **Working Principle of rear-control Hydraulic Impact Mechanisms**

Hydraulic impact mechanisms of this type are adopted primarily by the homemade hydraulic rock drills YYG250, YYG90A, and YYG100.

Figure 1.1 is the structural schematic of the domestically-produced YYG250 hydraulic rock drill. It is the first heavy-duty hydraulic rock drill developed in China for large-diameter drilling. Its maximum impact energy is 500 J, with an impact frequency of 32 Hz, and it is equipped with two high-pressure and two low-pressure accumulators. Besides the drilling rod rotation mechanism consisting of a hydraulic motor, gears, and transmission sleeves, and the slag discharge tube, all other components are the hydraulic impact mechanism. The key components are the cylinder body, impact piston, control valve, high-pressure accumulator, low-pressure accumulator, and the shank, which serves as the impacted body. The moving parts consist of the impact piston, control valve spool, and accumulator diaphragm. The YYG250 hydraulic rock drill adopts a rear-control working mode with a constant high pressure in the front chamber and alternating high and low pressures in the rear chamber. Refer to Fig. 1.2, the working principle is as follows:

(1) Return Stroke

In Fig. 1.2a, the piston has completed one impact, and the valve has reversed; The entire system is now at the beginning of the return stroke. At this point, the A-side of the valve spool is connected to the return oil (O) through oil passages 2 and 3, while the B-side is always connected to the low-pressure chamber 'a' of the valve body through the central hole of the spool. Additionally, the area of the A-side is larger than that of the B-side. Consequently, under the differential pressure of the high-pressure oil (P), the valve spool is positioned at the left end as shown in the

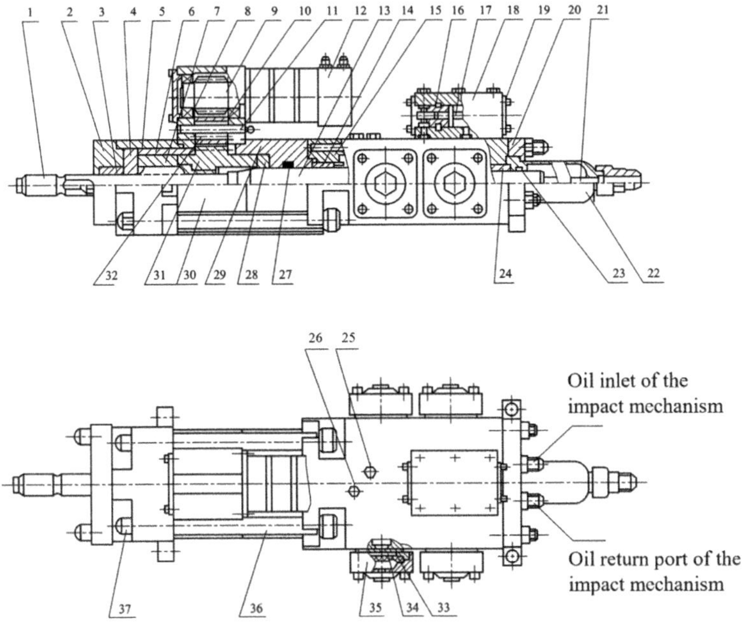

Fig. 1.1 Structure principle of domestic YYG250 hydraulic rock drill. 1- Shank; 2- Front cover; 3- Bushing; 4-Washer; 5- Front seat; 6- Bushing; 7- Transmission sleeve; 8- Bearing; 9- Gear; 10- Gear; 11-Shaft; 12- Hydraulic motor; 13- Piston; 14- Front sleeve; 15-Cylinder body; 16- Valve front sleeve; 17-Valve spool; 18- Valve body; 19- Valve rear sleeve; 20- Rear seat; 21-discharge tube; 22- Tail cover; 23- Seal; 24- Rear sleeve; 25- Plug; 26- Adjustment plug; 27- Seal; 28- Washer; 29- Intermediate body; 30- Gearbox; 31- Gear; 32- Bearing; 33- Accumulator diaphragm; 34-Accumulator lower cover; 35- Accumulator upper cover; 36- Long bolt; 37- Nut

diagram. The pressure oil (P) flows through the high-pressure chamber 'c' of the valve body, continuously maintains connecting with the piston's front chamber 'e'. Meanwhile, the piston's rear chamber 'f' is connected to the return oil through oil passage 4, and then through the control chamber 'b' and the low-pressure chamber 'a' of the valve body. As a result, the piston begins its return stroke to the left, driven by the pressure in the front chamber.

When the piston accelerates to the left, and its D-side crosses over the return control orifice 1, the A-side of the valve connects with the high-pressure oil (P) in the piston's front chamber through the orifice 1. Due to the high-pressure effect on the A-side, the valve spool starts its return stroke reverse motion to the right. The control chamber 'b' of the valve body connects with the high-pressure chamber 'c', and the front and rear chamber of the piston are differentially connected, both communicating with the high-pressure oil. This marks the end of the acceleration phase of the return stroke. Despite the fact that the F-side of the piston is larger than the D-side, the piston continues to move to the left due to inertia, but in deceleration until its velocity reaches zero, completing the entire return stroke. From the above,

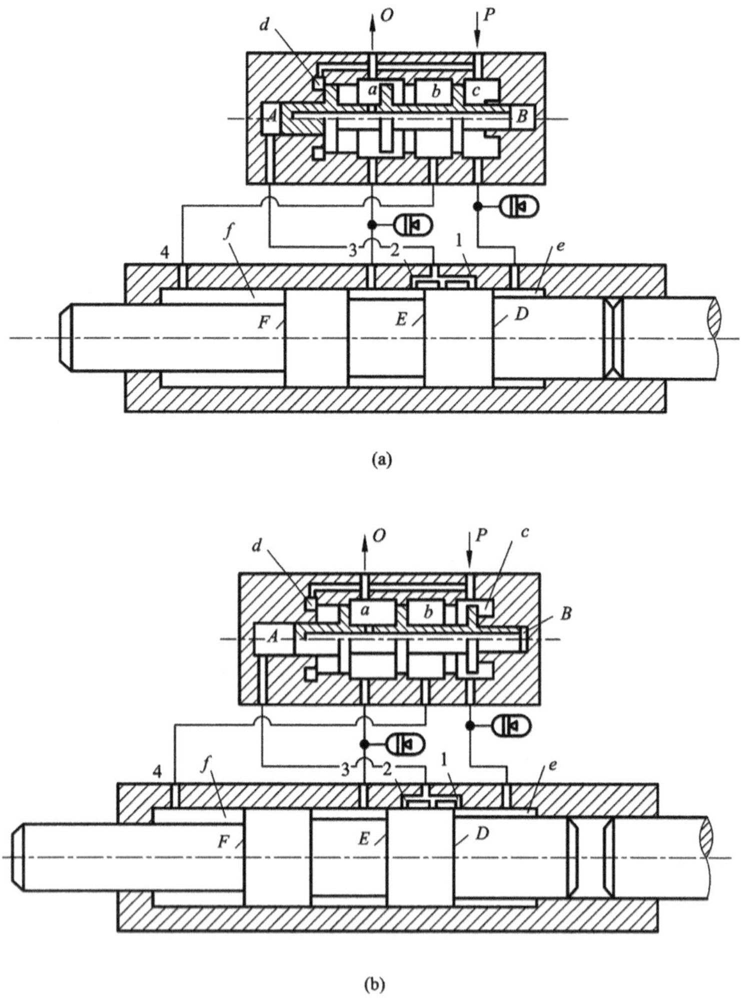

Fig. 1.2 Working principle of the impact mechanism of YYG250 hydraulic rock drill. **a** During return stroke phase **b** During impact stroke phase

it can be seen that the return stroke of the piston actually includes both acceleration and deceleration phases.

(2) Impact Stroke

After the completion of the return stroke, the system prepares for the impact stroke. At the beginning of the impact stroke, the overall oil circuit status is the same as during the return stroke deceleration phase, as shown in Fig. 1.2b. The valve spool remains on the right end, due to its A-side being connected to high-pressure oil. The front and rear chambers of the piston maintain their differential connection (both chambers are under high-pressure state). Under the action of pressure difference, the piston accelerates to the right, initiating the impact stroke.

When the E-side of the piston crosses the impact control orifice 2, the A-side of the valve spool is connected to the return oil through orifices 2 and 3, losing high-pressure oil and initiating the leftward motion for the impact stroke reversal. Subsequently, the piston impacts the shank, and the impact stroke comes to an end. The impact mechanism returns to the initial state of the return stroke, and the piston begins its return stroke again.

In this way, the piston continuously reciprocates, impacting the shank and delivering the crushing energy to break the rocks.

2. **Working Principle of dual-control Hydraulic Impact Mechanisms**

The hydraulic impact mechanism that adopts this control method mainly include the Cop1038 and Cop1238 hydraulic rock drills from Atlas-Copco company in Sweden.

The control principle of the domestic YYG80 hydraulic rock drill is shown in Fig. 1.3. It adopts a dual chamber control scheme with alternating high and low pressures in the front and rear chambers. The working principle is as below:

(1) Return Stroke

As shown in Fig. 1.3a, the piston has completed the impact, and the valve has completed the impact stroke reversal, and the entire system in the starting position of the return stroke. At this moment, the pressure oil (P) enters the front chamber ‘a’ of the piston through oil port 8, the high-pressure chamber ‘h’ and the front control chamber ‘k’ of the valve body, and then via oil passage 1. The rear chamber ‘c’ connects to the return oil (O) through oil passage 5, the rear control chamber ‘g’ and the low-pressure chamber ‘n’ of the valve body. The pressure in the front chamber drives the piston to the left, initiating the acceleration phase of the return stroke. When the piston reaches a certain position in the return stroke, the chamber ‘f’ of the valve connects to the high-pressure oil in the front chamber ‘a’ through the return stroke reversal control oil passage 2, while the chamber ‘e’ of the valve connects to the return oil (O) through oil passages 3 and 4. As a result, the valve spool moves to the left, causing the direction of the oil flow to reverse. The pressure oil (P) enters the rear chamber ‘c’ of the piston through the valve chambers ‘h’ and ‘g’ and oil passage 5, while the front chamber ‘a’ connects to the return oil (O) through oil passage 1, the front control chamber ‘k’ and another low-pressure chamber ‘q’ of the valve body. Due to inertia, the piston continues to move backward against the pressure in the rear chamber, gradually decelerating until its velocity reaches zero. The return stroke ends, and the impact stroke begins.

(2) Impact Stroke

After the completion of the return stroke, the system is prepared for the impact stroke. Fig. 1.3b shows the starting state of the impact mechanism’s impact stroke, at which point the rear chamber ‘c’ of the piston is connected to the high-pressure oil, and the front chamber ‘a’ is connected to the return oil. Under the pressure in the rear chamber, the piston starts to accelerate towards the right. Just before the impact

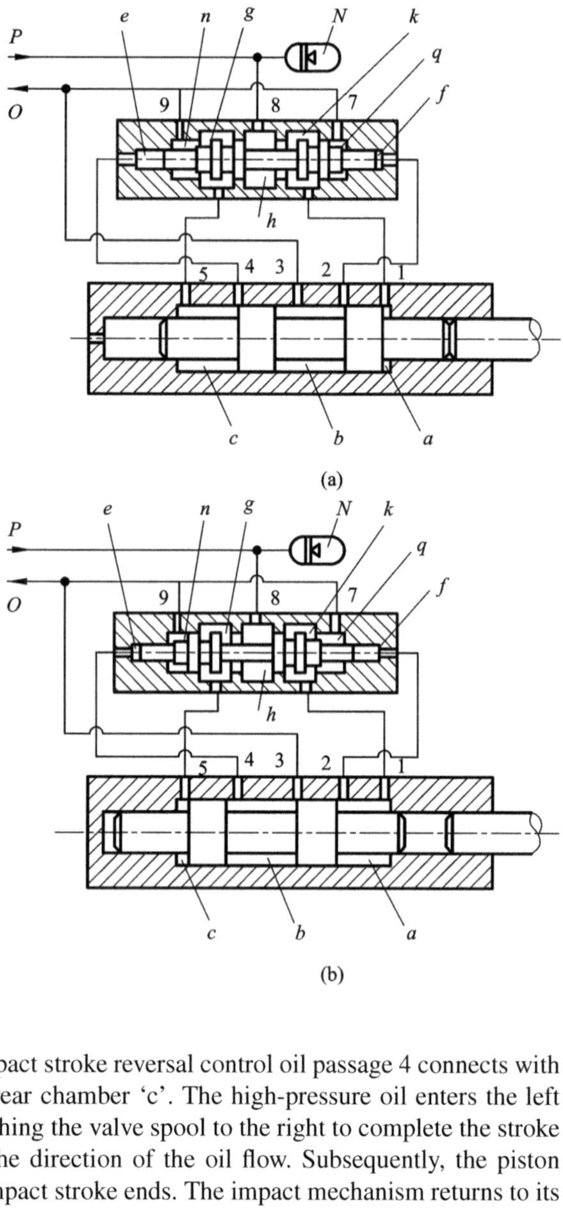

Fig. 1.3 The working principle of the impact mechanism of YYG80 hydraulic rock drill. **a** During return stroke phase **b** During impact stroke phase

stroke is about to end, the impact stroke reversal control oil passage 4 connects with the high-pressure oil in the rear chamber 'c'. The high-pressure oil enters the left chamber 'e' of the valve, pushing the valve spool to the right to complete the stroke reversal motion, switching the direction of the oil flow. Subsequently, the piston impacts the shank, and the impact stroke ends. The impact mechanism returns to its initial state for the return stroke, and the piston enters the return stroke of the next cycle.

In this way, the piston and the valve spool control each other through position feedback, continuously repeating the cycle, constantly impacting the shank, and outputting the impact energy.

1.3 Research Methods for Hydraulic Impact Mechanisms

1. Working Characteristics of Hydraulic Impact Mechanisms

Although hydraulic impact mechanisms have a simple structure, they belong to a relatively special type of hydraulic machinery. In comparison to general hydraulic machinery, they possess several prominent characteristics:

- During the operation of the impact mechanism, all moving components are constantly in a state of intense variable-speed motion, with acceleration of several tens or even hundreds of times higher than gravitational acceleration. It differs from the typical operation of other hydraulic machinery, which usually has a relatively stable working phase during the transition phases.
- The control valve is a high-frequency switching control valve with a switching frequency of 50-60 Hz. It is required to complete the switching action for a large flow rate within 1–2 ms. Although the servo valve also operates at a relatively high frequency, it always operates near the equilibrium point, and the fluid is always throttled.
- The above two characteristics will result in great inertia forces in the moving components of the impact mechanism. This inertia force is the primary factor caused by the working pressure of the transmission medium. In other words, the hydraulic pressure of the hydraulic impact mechanism depends mainly on the inertia forces of its moving components, which can be referred as inertia oil pressure. In general, it has little to do with the external load, i.e., the properties of the object being impacted and crushed. This is different from general hydraulic machinery, where the working oil pressure is mainly dependent on the external load, and it is relatively rare.
- Due to the intense variable-speed motion of the moving components in the impact mechanism, the flow state of the transmission medium that drives them naturally belongs to a very volatile non-steady flow. The pressure and flow rate changes are very drastic. Obviously, the inertial pressure generated by the inertial force of the hydraulic fluid flow cannot be ignored.
- The feedback control between the piston and the valve in the impact mechanism is not a definite mechanical feedback control; the control relationship between the piston and the valve is relatively free.

Obviously, the above-mentioned characteristics undoubtedly bring certain difficulties to the research, design, and manufacturing of hydraulic impact mechanisms. There are several challenging issues that need to be addressed, such as the establishment of mathematical models and the testing of parameters such as impact velocity and transient flow.

2. **Overview of Research Methods**

Researchers in the field of hydraulic impact mechanisms both domestically and internationally believe that the primary challenge in studying the design methodology for such impact mechanisms is how to describe the motion characteristics of the moving components.

The mathematical models describing the motion characteristics of hydraulic impact mechanisms can be categorized into two main types: linear systems of differential equations and nonlinear systems of differential equations. According to this, the existing theoretical research is classified into two categories in this book; it is referred as "linear model research" and "nonlinear model research."

(1) Study of the Linear Model of Hydraulic Impact Mechanisms

The laws of motion of the three main components of the impact mechanism are essentially nonlinear, and the mathematical model describing its motion should be a complex set of nonlinear differential equations. In the linear research method, the motion law is linearized in different ways, or some nonlinear effects are ignored, and the connection between the three moving parts is cut off to a certain extent, and then a relatively simple mathematical equation set is established to describe the motion of the piston.

The former Soviet union scholars, О.Д.Алимов and С.Абасов, were the first to point out that "under the condition of ensuring a given impact velocity, pressure control with equal oil pressure distribution is the most efficient and optimal control" (О.Д.Алимов, С.Абасов 1977). This control is represented in Fig. 1.4. The symbol 't1' represents the transition point (the end of the piston's return acceleration and the beginning of braking), which can be referred as the control point. 'F' represents the axial thrust acting on the piston, and 't' represents time. In this book, the control proposed by them is referred as "constant pressure control."

Research on linear models is essentially a research method based on the assumption of constant oil pressure.

(2) Nonlinear Model Research of Hydraulic Impact Mechanisms

The linear mathematical model of the impact mechanism mentioned above is a mathematical model of an ideal impactor under 'constant pressure control.' To achieve the ideal state of constant pressure, at least the following two basic conditions should be met:

Fig. 1.4 Constant pressure control model

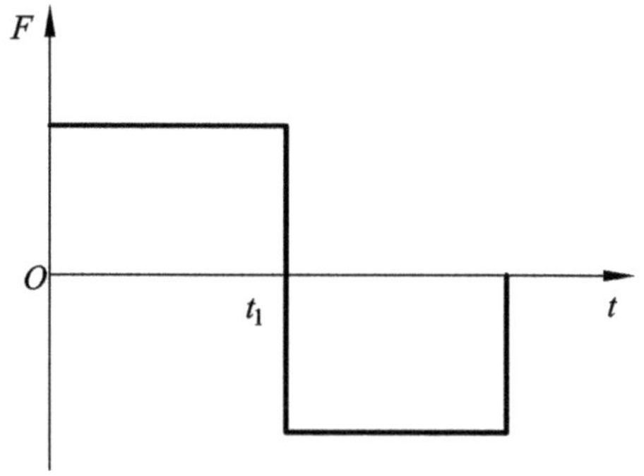

① The inflation volume of the accumulator is infinite.

If we define the inflation pressure ratio of the accumulator as $\alpha = p_a/p_{min}$, the working pressure ratio as $\gamma = p_{min}/p_{max}$, and ΔV as the maximum oil discharge volume of the accumulator in one working cycle, where p_a represents the accumulator inflation pressure, p_{min} represents the minimum working pressure of the accumulator, and p_{max} represents the maximum working pressure of the accumulator. According to the gas state equation $p_1 V_1^k = p_2 V_2^k$, the inflation volume of the accumulator V_a can be calculated.

$$V_a = \frac{\Delta V}{\alpha^{1/k} \cdot \left(1 - \gamma^{1/k}\right)}$$

The so-called constant pressure means that $p_{min} = p_{max} = Const$, therefore $\gamma = 1$. In the equation above, since ΔV is finite, when $\gamma \to 1, \ V_a \to \infty$.

The analysis above indicates that in order to maintain constant pressure, the inflation volume of the accumulator must be infinite.

② The duration of the spool reversing process is infinitesimal. Only in this way can occur the sudden pressure transition as shown in Fig. 1.4.

Obviously, these two conditions are practically unattainable. First of all, the working pressure of the accumulator generally follows a nonlinear variation as dictated by the adiabatic gas state equation. Similarly, the changes in valve opening result in nonlinear variations in pressure and flow rate. In order to make the mathematical model describing the motion law of the impact mechanism closer to the actual situation, researchers have started to establish nonlinear mathematical models for the impact mechanism.

The motion of the impact mechanism involves the combined movement of the piston, control valve, and accumulator. Moreover, this motion needs to be divided into several different stages according to the actual situation. This means that the boundary conditions and initial conditions for the differential equation system describing the motion of the impact mechanism become quite complex. Typically, numerical solutions are sought for such equation set. Therefore, computer numerical simulation techniques are indispensable for this nonlinear design and research.

(3) Characteristics and Comparison of the Two Research Methods

① The study of linear models is straightforward and easy to understand, and the design methods used allow for simplified calculations. Linear models have played a pioneering role in the research of hydraulic rock drills. In contrast, the study of nonlinear models is more complex and can only rely on computers for assisted research and design.

It is worth noting that the design methods of impact mechanisms guided by existing linear research are heavily reliant on empirical knowledge, with many details masked by rough empirical coefficients. As a result, the design of impact mechanisms almost becomes purely a technical issue. On the other hand, the design and research based on nonlinear mathematical models that better align with actual working conditions, combined with computer numerical simulation techniques, gradually transcend the realm of pure technicality and establish themselves on a more scientific theoretical foundation.

② The research of linear models can be used to obtain analytical expressions for the relationships between various parameters of the impact mechanism, and the analytical expressions can be applied conveniently in guided research. But generally, nonlinear model research cannot obtain analytical expressions. It only can be carried out through computer-based analytical approaches, which means that the research should be carried out by the simulation process of computers.

③ The motion of the impact mechanism is determined by the combined motion of its three major components. However, the study of linear models does not simultaneously reflect the connection and constraints between these three components, nor can it capture the relationships among numerous other structural parameters within the machine. Therefore, from a certain perspective, linear model research can be considered a static, isolated, and idealized research method. On the other hand, although nonlinear model research may not provide analytical expressions for the relationships between various parameters, it establishes a mathematical model that comprehensively considers the complex physical processes of the combined motion of the three major components and the numerous structural parameters of the impact mechanism. As a result, the simulation model running on a computer can more accurately reflect the motion characteristics of the impact mechanism. Consequently, nonlinear model research represents a dynamic, comprehensive, and more realistic research approach.

④ As mentioned earlier, linear model research is an idealized research method. However, studying ideal models in scientific research is necessary because they often provide directional insights and indicate the targets that actual models should strive to approximate. Generally speaking, linear model research cannot be used for more precise design-oriented studies, but this limitation can be compensated by the strengths of nonlinear research. After a well-developed simulation model is executed, it can yield a set of relatively accurate results with certain predictive capabilities, presenting the motion processes of the actual mechanism. One can envision the following scenario: the linear model first establishes a rough optimal constraint range, and then the input parameters of the simulation model are modified. After several "experiments," a specific point within this constraint range is determined. In this way, these two research methods can complement each other. This represents the fundamental approach of computer-aided design for hydraulic impact mechanisms.

References

О.Д.Алимов, С.Абасов. (1977). Theory of hydraulic vibration impact mechanism. In *Mining and Metallurgy Translations (Mining)* (in Chinese). Changsha: Changsha Mining Research Institute of the Ministry of Metallurgical Industry.

Cao, Z. C., & Zi, Y. (1989). Overview of the development of hydraulic rock drills and jumbos. *World Mining Express*, (35), 20–22,16. https://doi.org/10.13828/j.cnki.ckjs.1989.35.011.

Guo, R. L. (1990). Special lecture on hydraulic rock drill (Part 1). *Rock Drilling Machinery & Pneumatic Tools* (in Chinese), (3), 67–73. https://next.cnki.net/middle/abstract?v=r9IaLYgXogUmGDN0kCQ7aZWCIpteDUPL34UpXhsF7Wo5fhhAVO_ThSvzsCzGynaNO4KC52umVVpcm3kDlmc4tr_lWoWkMBL26IBNe6JVru8iWgAvQHqF5GUf59NW9UUqznsyAiIABPO4jhNe5KY7N9rhwwdRptoTrMiatF7iqI1qAB_pZeCsPm95EvhmxCRR&uniplatform=NZKPT&language=CHS&scence=null.

Guo, X. X. (1993). Foreign hydraulic rock drilling equipment. *Geology Press* (in Chinese).

He, Q. H., Yang, X. B., & Xia, J. S. (1990). Development of hydraulic rock drilling equipment for open-pit mining. *Mining Machinery* (in Chinese), (3), 7–10. https://next.cnki.net/middle/abstract?v=th5-mUcNE0NSxxnsJ1kNlfEEAGfpIod98cOS8LhFhdG0VHHXdMdHYqFnJW_V1Ps0p2imDHiIY7_c7VshXciLKthg7pfW5SYklAwl56T981kEteW-NeRy4ZZW9xCuETt6ZVDEILf5G4KAZ2Z74cL2WHnFXIB7GqWu3svlY4Ykv_-RdNHUgUUrsnYpdkTH8P9k&uniplatform=NZKPT&language=CHS&scence=null

Jiang, Z. J. (1995). Application of hydraulic rock drilling equipment in underground mines in China. *Mining Research and Development* (in Chinese), (S1), 81–84. https://next.cnki.net/middle/abstract?v=amOBmv6QLtoD6sAokRFP_IoREHU-5EH2m5LeDwz4w4jSbGetIpSh4YGs2oPEfE3_wEyKm8Yd9A2dj2RAL4LGZMjDei5pPwC083BisbTgbq9V6QVdethyq44hRlDd-rbgMDj8VLN617e32oEFSlIpsGW78eAv__oBDsoWi7ELD2N2DVSLiGZQ-Wjh3Hdlyv4m&uniplatform=NZKPT&language=CHS&scence=null

Li, Y. Q. (1991). Current status and development of rock drilling machinery in China. *Mining & Processing Equipment* (in Chinese), (7), 29–33. https://next.cnki.net/middle/abstract?v=amOBmv6QLtoMvO1kCIsu8kv-ndy6HJUIU5fZcgA7pymP1Y1hkteOfxp13_pvEXuzHLD0Fie71RXOtcthIskzdZxBWuLsnCkFAw0a3torVAXAE0mm6BzKiBph04Lv5bzb8C4b0cEgROmmUQZsVkVPYV7FZFFPa_54DD7bPYONqZRy0vB4EmR3eSNq6I-AKfh8&uniplatform=NZKPT&language=CHS&scence=null

Wang, W. H. et al. (1987). Foreign hydraulic rock drills. *China Coal Industry Publishing House* (in Chinese).

Xia, J. S. (1985). Development of fully hydraulic rock drilling equipment abroad. *Mining & Processing Equipment* (in Chinese), (8), 54–58. https://next.cnki.net/middle/abstract?v=amOBmv6QLtraVLAaVx5ekRtSTA9ZGjJGN7DoWKaMrnjmyV2nEQ8vsE5hmjwHaoSptxtdQexz7yc0F8KiIEGxLOhmcUyLrniG021FL29rMlohIkbRY0rFucow6T5ZRyowI50nHzOj7aDmr8ezSo0Ko0mOQhF2JSjVZ1S0zOC8becvLbJuj1-YrXWKK8bwYGle&uniplatform=NZKPT&language=CHS&scence=null

Xia, J. S. (1987). Development of hydraulic rock drilling equipment and its application in open-pit mining (rock engineering). *Rock Drilling Machinery & Pneumatic Tools* (in Chinese), (3), 29–45. https://next.cnki.net/middle/abstract?v=amOBmv6QLtpOHHqs77y5SRcYLuaBB9dpYK9qnDwlRVxBYSstvH9kzzF1ailCW9i4B8--EX9uaHN1DA1Hk8X2fwB_BI6mE-5s10r9cgojecFzQ5I5OLFALZq_UENrzIws17x5FYZwJAqliMwHLAkM6_H2TirRlYgibX6RrgPkJnAB7RxDY3FIkujzFp_7cFMi&uniplatform=NZKPT&language=CHS&scence=null

Xia, J. S. (1990). Research, development, and promotion of hydraulic rock drilling equipment in China. *Rock Drilling Machinery & Pneumatic Tools* (in Chinese), (4), 23–33. https://next.cnki.net/middle/abstract?v=amOBmv6QLtqZFZkygFCMkLW5_8BzpzQ02kACQ2F2fXYtFWBaJW9ih9TVHWOm_bHu1s4H3xjNVsxRzzwJzXqqfbBfvd73tvGjML3x2MhkA37NWTqEWSOek9NeDnePkoBj9VD9bJbD0CkHv_sgh6H3DjnQ69I4xKjqWsl9adgWLFdikH2h3aBT6cAlwnb0Xtxz&uniplatform=NZKPT&language=CHS&scence=null

Xia, J. S. (1993). Progress in the development and application of hydraulic rock drilling equipment. *Rock Drilling Machinery & Pneumatic Tools* (in Chinese), (2), 11–18, 70. https://next.cnki.net/middle/abstract?v=amOBmv6QLtqfyiF5ejLhhbthorGiEX54DWTrEpfmTPZm3xswmwi5CgUyrYVmCmwlNcfy2lX0bmsvfhrjjFDTidTilxnW7hafaWy32LO42JHJch3s_SfTEDazmhrW7KWCgZpwFRwuL0nxi1ByKO37OGhT_MVvedcEYdKiujBQ6S4xDd-f5SKMZL-4TxNk99lH&uniplatform=NZKPT&language=CHS&scence=null

Zhang, G. Z. (1991). Pneumatic impact equipment and its design. *China Machine Press* (in Chinese).

Zhao, T. W. (1996). Dynamics of percussive drilling. *Metallurgical Industry Press* (in Chinese).

Chapter 2
Three-Step Analysis of the Piston Movement of Hydraulic Impact Mechanisms

2.1 Parameters and Symbols

In this book, various parameters representing the hydraulic impact mechanism are categorized into three major classes with corresponding symbols.

1. **Performance Parameters**

These parameters directly reflect the performance indicators of the impact mechanisms.

- Impact Frequency (f)
- Impact Energy (E_i)
- Overall Efficiency (η)
- Volumetric Efficiency (η_v)
- Mechanical Efficiency (η_m)
- Pressure Efficiency (η_p)

Among these, Pressure Efficiency primarily reflects the magnitude of pressure losses caused by local obstructions in the internal oil passages of the hydraulic impact mechanisms.

2. **Structural parameters**

Parameters related to the structure of the impact mechanism.

- Piston Mass (m_p)
- Effective Action Area of the Front and Rear chambers of the piston (A_2, A_1)
- Accumulator Inflation Volume (V_a)

3. Working Parameters

Parameters that describe the working condition of the impact mechanism.

- Working Oil Pressure (p_i)

Q. He, *Research and Design of Hydraulic Impact Mechanisms*,
https://doi.org/10.1007/978-981-96-4864-1_2

- Input Flow Rate (Q_i)
- Leakage Flow Rate (Q_l)
- Piston Characteristic Speeds: This includes the impact velocity (u_{im}), maximum velocity during return acceleration phase (u_{rm}).
- Piston Stroke (S_p): It can be further divided into return acceleration stroke (S_r) and return deceleration stroke (S_d).
- Piston Movement Cycle (T): Including the return acceleration time (t_r), return deceleration time (t_d), and impact stroke time (t_i).
- Piston Accelerations: Including the return acceleration phase acceleration (ω_r), return deceleration phase acceleration (ω_d), and impact stroke acceleration (ω_i).
- Accumulator Parameters: Including accumulator inflation pressure (P_a), the oil charging volume during the return acceleration phase (ΔV_{ri}), oil discharge volume during the return acceleration phase (ΔV_{ro}), and the oil volumes of charging (ΔV_i) and discharging (ΔV_o) during the return deceleration and impact phases.

2.2 Kinematic Analysis of the Piston

Given that the supply pressure to the piston is assumed to be equal during the return deceleration and impact acceleration phases, and that the areas subjected to hydraulic pressure are also the same, why is there a difference in the acceleration values between these two states? The main reason is due to the sealing resistance, viscous friction resistance, and hydraulic clamping force, which create the same deceleration resistance as the oil pressure during the return deceleration phase. However, in the impact acceleration phase, these forces oppose the direction of the oil pressure that drives the piston's acceleration. Clearly, this will result in the negative acceleration during the return deceleration phase being greater in magnitude than the acceleration during the impact phase. For ease of analysis, two dimensionless variables are introduced in this book: the ratio of acceleration between the return acceleration phase and the impact acceleration phase, donated as $\beta = \frac{\omega_r}{\omega_i}$, referred to as the return-impact acceleration ratio, and the ratio of acceleration values (numerical) between the impact acceleration phase and the return deceleration phase, donated as $\beta_1 = \frac{\omega_i}{\omega_d}$, referred to as the impact-deceleration acceleration ratio, as shown in Fig. 2.1.

For a given impact frequency f, impact energy E_i, with an appropriate impact velocity u_{im}, input oil pressure P_i, and the ratio β, a series of formulas can be derived to describe the kinematic relationships of the piston in the impact mechanism:

$$m_p = \frac{2E_i}{u_{im}^2} \tag{2.1}$$

$$T = \frac{1}{f} \tag{2.2}$$

If the impact pause time is ignored, we have

Fig. 2.1 Velocity and flow variation of the piston in a hydraulic impact mechanism

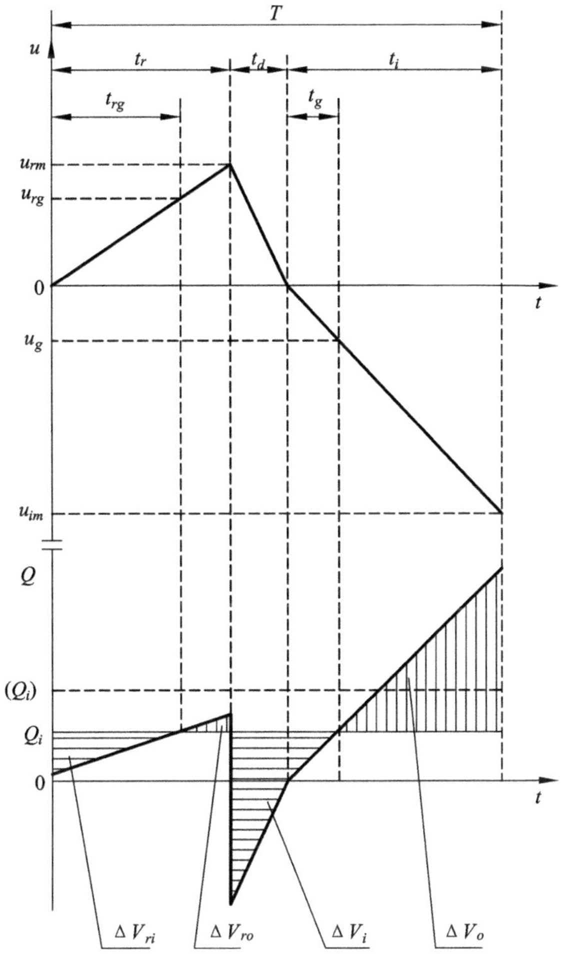

$$T = t_r + t_d + t_i, \tag{2.3}$$

where

$$t_i = \frac{2S_p}{u_{im}}, \tag{2.4}$$

$$t_r = \frac{u_{rm}}{\omega_r} = \frac{u_{rm}}{\beta\beta_1\omega_d} = \frac{t_d}{\beta\beta_1} \tag{2.5}$$

In addition:

$$S_r = \frac{\omega_r t_r^2}{2}, \tag{2.6}$$

$$S_d = \frac{\omega_d t_d^2}{2}, \tag{2.7}$$

$$S_p = S_r + S_d = \frac{u_{im}^2}{2\omega_i}. \tag{2.8}$$

Solving the above formula simultaneously yields

$$S_d = \frac{\beta\beta_1}{1+\beta\beta_1} S_p, \tag{2.9}$$

$$S_r = \frac{1}{1+\beta\beta_1} S_p, \tag{2.10}$$

$$t_d = 2\beta_1 \sqrt{\frac{\beta}{1+\beta\beta_1}} \cdot \frac{S_p}{u_{im}}, \tag{2.11}$$

$$t_r = \frac{2}{\beta} \sqrt{\frac{\beta}{1+\beta\beta_1}} \cdot \frac{S_p}{u_{im}}, \tag{2.12}$$

$$T = \frac{2[\beta + \sqrt{\beta(1+\beta\beta_1)}]}{\beta} \cdot \frac{S_p}{u_{im}}. \tag{2.13}$$

Let

$$\psi_{sp} = \frac{\beta}{2[\beta + \sqrt{\beta(1+\beta\beta_1)}]}, \tag{2.14}$$

$$\psi_{ur} = \sqrt{\frac{\beta}{(1+\beta\beta_1)}}. \tag{2.15}$$

Further useful formula can be obtained:

$$S_p = \psi_{sp} u_{im} T, \tag{2.16}$$

$$S_d = \beta_1 \psi_{ur}^2 S_p = \beta_1 \psi_{ur}^2 \psi_{sp} u_{im} T, \tag{2.17}$$

$$S_r = \beta^{-1} \psi_{ur}^2 \psi_{sp} u_{im} T, \tag{2.18}$$

$$\omega_i = \frac{u_{im}}{2\psi_{sp} T}, \tag{2.19}$$

$$\omega_r = \beta\omega_i = \frac{\beta u_{im}}{2\psi_{sp} T}, \tag{2.20}$$

$$\omega_d = \frac{\omega_i}{\beta_1} = \frac{u_{im}}{2\beta_1\psi_{sp}T}, \tag{2.21}$$

$$u_{rm} = \psi_{ur}u_{im}, \tag{2.22}$$

$$t_i = 2\psi_{sp}T, \tag{2.23}$$

$$t_d = 2\beta_1\psi_{ur}\psi_{sp}T, \tag{2.24}$$

$$t_r = 2\beta^{-1}\psi_{ur}\psi_{sp}T. \tag{2.25}$$

Under the ideal condition of $\beta_1 = 1$

$$\psi_{sp} = \frac{\sqrt{\beta(1+\beta)} - \beta}{2}, \tag{2.26}$$

$$\psi_{ur} = \sqrt{\frac{\beta}{(1+\beta)}}. \tag{2.27}$$

Let's assume that the time required for the control valve to switch oil circuits during the return stroke and the impact stroke is denoted as t_v and t_{iv}, respectively. For convenience of calculation, let's define $t_v = k_v t_r$ and $t_{iv} = k_{iv} t_i$. Here k_v and k_{iv} are the return stroke reversing time coefficient and the impact stroke reversing time coefficient, respectively. Then, we can obtain:

The position of the return control oil port (i.e., the distance from the start of the piston's return stroke to the point where it opens the oil port that activates the control valve reversal) is:

$$\begin{aligned} S_c &= \frac{\omega_r(t_r - t_v)^2}{2} = \frac{\omega_r(1-k_v)^2 t_r^2}{2} \\ &= (1-k_v)^2 S_r. \end{aligned} \tag{2.28}$$

The position of the impact stroke control oil port (i.e., the distance from the start of the piston's impact stroke to the point where it opens the oil port that activates the control valve reversal) is:

$$\begin{aligned} S_{ic} &= \frac{\omega_i(t_i - t_{iv})^2}{2} = \frac{\omega_i(1-k_{iv})^2 t_i^2}{2} \\ &= (1-k_{iv})^2 S_p. \end{aligned} \tag{2.29}$$

The formulates (2.16) to (2.27) indicate that the working parameters related to the kinematics of the piston of the impact mechanism can be expressed by the impact period (frequency), impact velocity (energy) and two acceleration ratios, independent of the magnitude of the input working oil pressure.

2.3 Front-Chamber Area (A_2) and Rear-Chamber Area (A_1)

1. **Coefficient of Resistance**

The hydraulic impact mechanism is subject to various resistances during its motion. From a practical perspective, the following two main resistance coefficients are considered:

(1) Return Oil Resistance Coefficient k_0

For the convenience of analysis, the return oil pressure p_0 is represented in terms of the working oil pressure p_i:

$$p_0 = k_0 p_i \tag{2.30}$$

Typically, $k_0 = 0.06 \sim 0.12$.

(2) Comprehensive Resistance Coefficient k_y

In addition to reflecting the sealing resistance, it also reflects the viscous friction resistance and the hydraulic locking resistance. To facilitate the establishment of a linear model, the resistance pressure p_y generated by these resistances is also expressed with the working pressure p_i:

$$p_y = k_y p_i \tag{2.31}$$

Typically, $k_y = 0.05 - 0.1$.

The expressions for the oil pressure acting area of the front and rear chambers of the piston, as well as impact-deceleration acceleration ratio β_1, will be further obtained for the rear-control type and the dual-control type of the hydraulic impact mechanism.

2. **A_2, A_1 and β_1 of the Rear-control Type**

The acceleration expressions for each state can be derived based on the forces acting on the piston:

$$\begin{aligned} \omega_r &= \frac{p_i A_2 - p_y A_2 - p_0 A_1}{m_p} \\ &= \frac{\left[(1 - k_y)A_2 - k_0 A_1\right] p_i}{m_p}, \end{aligned} \tag{2.32}$$

$$\omega_d = \frac{p_i(A_1 - A_2) + p_y(A_1 + A_2)}{m_p} = \frac{\left[(1 + k_y)A_1 - (1 - k_y)A_2\right]p_i}{m_p}, \tag{2.33}$$

$$\omega_i = \frac{p_i(A_1 - A_2) - p_y(A_1 + A_2)}{m_p} = \frac{\left[(1 - k_y)A_1 - (1 + k_y)A_2\right]p_i}{m_p}. \tag{2.34}$$

By solving Eqs. (2.32), (2.34), and (2.19) simultaneously, it can be obtained

$$A_1 = \frac{\left[(1 - k_y) + (1 + k_y)\beta\right]m_p\omega_i}{\left[(1 - k_y)^2 - k_0(1 + k_y)\right]p_i} = \frac{\left[(1 - k_y) + (1 + k_y)\beta\right]m_p u_{im}}{2\psi_{sp}\left[(1 - k_y)^2 - k_0(1 + k_y)\right]p_i T} = \psi_{A1}\frac{m_p u_{im}}{p_i T} \tag{2.35}$$

$$A_2 = \frac{\left[k_0 + (1 - k_y)\beta\right]m_p\omega_i}{\left[(1 - k_y)^2 - k_0(1 + k_y)\right]p_i} = \frac{\left[k_0 + (1 + k_y)\beta\right]m_p u_{im}}{2\psi_{sp}\left[(1 - k_y)^2 - k_0(1 + k_y)\right]p_i T} = \psi_{A2}\frac{m_p u_{im}}{p_i T} \tag{2.36}$$

How is the other acceleration ratio β_1 determined?

In fact, after selecting β, k_0, and k_y, and substituting Eqs. (2.35) and (2.36) into Eq. (2.33), another acceleration ratio can be obtained

$$\beta_1 = \frac{(1 - k_y)^2 - k_0(1 + k_y)}{(1 - k_y)(1 + k_y - k_0) + 4k_y\beta}. \tag{2.37}$$

Conversely, on the existing A_1 and A_2, Eqs. (2.35) and (2.36) can be used to obtain

$$\beta = \frac{(1 - k_y)A_2 - k_0 A_1}{(1 - k_y)A_1 - (1 + k_y)A_2}, \tag{2.38}$$

$$A_2 = \beta_2 A_1, \tag{2.39}$$

where

$$\beta_2 = \frac{k_0 + (1 - k_y)\beta}{1 - k_y + (1 + k_y)\beta}. \tag{2.40}$$

3. **A_2, A_1 and β_1 of the Dual-control Type**

Due to the alternating pressure in the front and rear chambers, with no constant-pressure chamber, the expressions for the magnitude of the acceleration become:

$$\omega_r = \frac{[(1 - k_y)A_2 - k_0 A_1]p_i}{m_p}, \tag{2.41}$$

$$\omega_d = \frac{[(1 + k_y)A_1 - k_0 A_2]p_i}{m_p}, \tag{2.42}$$

$$\omega_i = \frac{[(1 - k_y)A_1 - k_0 A_2]p_i}{m_p}. \tag{2.43}$$

Similarly, it can be obtained

$$\begin{aligned} A_1 &= \frac{[(1 - k_y) + k_0\beta]m_p\omega_i}{[(1 - k_y)^2 - k_0^2]p_i} \\ &= \frac{[(1 - k_y) + k_0\beta]m_p u_{im}}{2\psi_{sp}[(1 - k_y)^2 - k_0^2]p_i T} \\ &= \psi_{A1}\frac{m_p u_{im}}{p_i T}, \end{aligned} \tag{2.44}$$

$$\begin{aligned} A_2 &= \frac{[(1 - k_y)\beta + k_0]m_p\omega_i}{[(1 - k_y)^2 - k_0^2]p_i} \\ &= \frac{[(1 - k_y)\beta + k_0]m_p u_{im}}{2\psi_{sp}[(1 - k_y)^2 - k_0^2]p_i T} \\ &= \psi_{A2}\frac{m_p u_{im}}{p_i T}, \end{aligned} \tag{2.45}$$

$$\beta_1 = \frac{(1 - k_y)^2 - k_0^2}{(1 - k_y)(1 + k_y) + 2k_0 k_y \beta - k_0^2}, \tag{2.46}$$

$$\beta = \frac{(1 - k_y)A_2 - k_0 A_1}{(1 - k_y)A_1 - k_0 A_2}, \tag{2.47}$$

$$A_2 = \beta_2 A_1, \tag{2.48}$$

where

$$\beta_2 = \frac{k_0 + (1 - k_y)\beta}{1 - k_y + k_0\beta}. \tag{2.49}$$

It should also be noted that, under the ideal no-resistance conditions of $k_0 = k_y = 0$ and $\beta_1 = 1$, the return-impact acceleration ratio β is equal to the ratio of the effective oil pressure areas in the front and rear chambers of the piston. In this case, we have

$$\beta = \left[\begin{array}{ll} \frac{A_2}{A_1 - A_2} & \text{(the rear - control type)} \\ \frac{A_2}{A_1} & \text{(the dual - control type)} \end{array}\right., \tag{2.50}$$

$$\beta_2 = \left[\begin{array}{ll} \frac{\beta}{1+\beta} & \text{(the rear - control type)} \\ \beta & \text{(the dual - control type)} \end{array}\right.. \tag{2.51}$$

Additionally, several other coefficients are simplified as:

$$\psi_{A1} = \left[\begin{array}{ll} \frac{1+\beta}{2\psi_{sp}} & \text{(the rear - control type)} \\ \frac{1}{2\psi_{sp}} & \text{(the dual - control type)} \end{array}\right., \tag{2.52}$$

$$\psi_{A2} = \left[\begin{array}{ll} \frac{\beta}{2\psi_{sp}} & \text{(the rear - control type)} \\ \frac{\beta}{2\psi_{sp}} & \text{(the dual - control type)} \end{array}\right.. \tag{2.53}$$

2.4 Input Flow Rate Q_i

During one impact cycle, besides the control valve, the high-pressure oil is primarily consumed in the following aspects:

1. **Effective Oil Volume V_e**

As is well known in the industry, accumulators are generally equipped in hydraulic impact mechanisms to recover the kinetic energy possessed by the piston during its return stroke to compensate for insufficient flow rate during impact. Therefore, under ideal conditions, the effective oil volume is the hydraulic oil consumed during the piston's movement in the impact stroke. The consumption value within one cycle is

$$V_e = \left[\begin{array}{ll} (A_1 - A_2)S_p & \text{(the rear - control type)} \\ A_1 S_p & \text{(the dual - control type)} \end{array}\right.. \tag{2.54}$$

V_e represents the effective volume of high-pressure oil consumed by the impact mechanism.

And $\frac{V_e}{T}$ is referred as the effective flow rate Q_e.

2. **Leakage Oil Volume V_l**

Using the flow rate formula for an eccentric circular ring gap and based on the dimensions of common piston gaps at the front, rear, and intermediate three-stage seals, the leakage calculation coefficients for the front and rear chambers, k_{2l} and k_{1l}, can be determined sequentially. Then, according to the leakage time in the pressure chamber, the pressure oil loss due to leakage over one cycle can be calculated.

$$\begin{aligned} V_l &= \left[\begin{array}{l} [k_{2l}T + k_{1l}(t_d + t_i)]p_i \\ [k_{2l}t_r + k_{1l}(t_d + t_i)]p_i \end{array}\right. \\ &= \left[\begin{array}{ll} \left[k_{2l} + 2k_{1l}\psi_{sp}(1 + \beta_1\psi_{ur})\right]p_i T & \text{(the rear - control type)} \\ 2\psi_{sp}\left[\frac{k_{2l}\psi_{ur}}{\beta} + k_{1l}(1 + \beta_1\psi_{ur})\right]p_i T & \text{(the dual - control type)} \end{array}\right. \\ &= \varphi_l p_i T, \end{aligned} \tag{2.55}$$

where

$$\varphi_l = \left[\begin{array}{l} k_{2l} + 2k_{1l}\psi_{sp}(1 + \beta_1\psi_{ur})\text{(the rear - control type)} \\ 2\psi_{sp}[\frac{k_{2l}\psi_{ur}}{\beta} + k_{1l}(1 + \beta_1\psi_{ur})\text{(the dual - control type)} \end{array}\right. .$$

3. **Stroke Loss Oil Volume V_u**

The high-pressure oil volume consumed by the piston during the return acceleration phase, $V_r = A_2S_r$ is converted into the kinetic energy of the piston. This part of the energy can be stored in the accumulator during the return deceleration phase by injecting high-pressure oil with a volume of $V_d = (A_1 - A_2)S_d$ (for rear-control type) or $V_d = A_1S_d$ (for dual-control type) into it. Under ideal conditions, all the energy from the return stroke is fully recovered. However, in actual working conditions, $k_0 \neq 0$, $k_y \neq 0$, i.e. $\beta_1 < 1$, the value of S_d calculated with formula (2–17) during return deceleration stroke is smaller than the value calculated under the ideal working condition where $k_0 = k_y = 0$, i.e. $\beta_1 = 1$ (see the relevant analysis in the next chapter for details). In other words, the volume of recovered oil is less than the volume of consumed oil, that is to say $V_d < V_r$.

Since this oil volume loss is related to the reduction in the return braking stroke, we refer to it simply as the "stroke loss oil volume." The volume of high-pressure oil consumed due to stroke loss in one cycle is:

$$\begin{aligned} V_u = V_r - V_d &= \left[\begin{array}{l} A_2S_p - A_1S_d \\ A_2S_r - A_1S_d \end{array}\right. \\ &= \left[\begin{array}{ll} \frac{2\psi_{A1}\psi_{sp}(\beta_2 - \beta_1\psi_{ur}^2)E_i}{p_i} & \text{(the rear - control type)} \\ \frac{2\psi_{A1}\psi_{sp}\psi_{ur}^2\left(\frac{\beta_2}{\beta} - \beta_1\right)E_i}{p_i} & \text{(the dual - control type)} \end{array}\right. \end{aligned} \tag{2.56}$$

The flow rate $Q_u = \frac{V_u}{T}$ is referred as the stroke loss flow rate.

4. **Input Flow Rate Q_i**

Based on the above analysis, the expression for the actual input flow rate is

$$Q_i = \frac{V_e + V_l + Vu}{T} = \left[\begin{array}{l} \frac{A_1 S_r + \varphi_l p_i T}{T} \\ \frac{(A_1 + A_2) S_r + \varphi_l P_i T}{T} \end{array} \right. = \varphi_l p_i + \frac{\varphi_q E_i}{p_i T}. \tag{2.57}$$

where

$$\varphi_q = \left[\begin{array}{ll} \frac{2\psi_{A1}\psi_{sp}\psi_{ur}^2}{\beta} & (the\ rear - control\ type) \\ \frac{2(1+\beta_2)\psi_{A1}\psi_{sp}\psi_{ur}^2}{\beta} & (the\ dual - control\ type) \end{array} \right. . \tag{2.58}$$

2.5 Charging and Discharging Oil Volume, Inflation Volume, and Inflation Pressure of High-Pressure Accumulator

In general, the high-pressure accumulator undergoes twice charging and twice discharging cycles within one working cycle, as shown in Fig. 2.1, and it is necessary to analyze their variation for the research and design of accumulators. From Fig. 2.1, it can be observed that the moment t_{rg} and t_g, at which the flow rate required by the front and rear chambers during piston movement is equal to the supplied oil flow rate after deducting the leakage flow rate ($Q_{eu} = Q_e + Q_u$), serve as the turning points between the charging and discharging phases. By assuming the piston velocity at these moments as u_{rg} and u_g, the following calculation formula can be obtained.

Since

$$\begin{aligned} Q_{eu} &= \frac{A_1 S_r}{T} \\ &= \left[\begin{array}{rl} A_2 u_{rg} &= A_2 \omega_r t_{rg} \\ (A_1 - A_2) u_g &= (A_1 - A_2) \omega_i t_g \end{array} \right. \text{(the rear - control type)}, \end{aligned} \tag{2.59}$$

and

$$\begin{aligned} Q_{eu} &= \frac{(A_1 + A_2) S_r}{T} \\ &= \left[\begin{array}{rl} A_2 u_{rg} &= A_2 \omega_r t_{rg} \\ A_1 u_g &= A_1 \omega_i t_g \end{array} \right. \text{(the dual - control type)}. \end{aligned} \tag{2.60}$$

So, we can get

$$t_{rg} = \left[\begin{array}{ll} \frac{2\psi_{ur}^2 \psi_{sp}^2}{\beta_2 \beta^2} \cdot T & (the\ rear - control\ type) \\ \frac{2(1+\beta_2)\psi_{ur}^2 \psi_{sp}^2}{\beta_2 \beta^2} \cdot T & (the\ dual - control\ type) \end{array} \right., \tag{2.61}$$

$$t_{g} = \left[\begin{array}{ll} \frac{2\psi_{ur}^2 \psi_{sp}^2}{(1-\beta_2) \beta} \cdot T & (the\ rear - control\ type) \\ \frac{2(1+\beta_2)\psi_{ur}^2 \psi_{sp}^2}{\beta} \cdot T & (the\ dual - control\ type) \end{array} \right.. \tag{2.62}$$

Referencing the area of the four shaded regions in the flow rate graph in Fig. 2.1, the charging and discharging volumes of the accumulator can be determined.

$$\Delta V_{ri} = \frac{Q_{eu} t_{rg}}{2} = \left[\begin{array}{ll} \frac{2\psi_{A1} \psi_{ur}^4 \psi_{sp}^2}{\beta_2 \beta^3} \cdot \frac{E_i}{p_i} & (the\ rear - control\ type) \\ \frac{2(1+\beta_2)^2 \psi_{A1} \psi_{ur}^4 \psi_{sp}^2}{\beta_2 \beta^3} \cdot \frac{E_i}{p_i} & (the\ dual - control\ type) \end{array} \right., \tag{2.63}$$

$$\begin{aligned} \Delta V_{r0} &= \frac{(A_2 u_{rm} - Q_{eu})(t_r - t_{rg})}{2} \\ &= \left[\begin{array}{ll} \frac{2\beta_2 \psi_{A1} \psi_{ur}^2 \psi_{sp}}{\beta} \left(1 - \frac{\psi_{ur}\psi_{sp}}{\beta_2 \beta}\right)^2 \frac{E_i}{p_i} & (the\ rear - control\ type) \\ \frac{2\beta_2 \psi_{A1} \psi_{ur}^2 \psi_{sp}}{\beta} \left[1 - \frac{(1+\beta_2)\psi_{ur}\psi_{sp}}{\beta_2 \beta}\right]^2 \frac{E_i}{p_i} & (the\ dual - control\ type) \end{array} \right., \end{aligned} \tag{2.64}$$

$$\begin{aligned} \Delta V_i &= \left[\begin{array}{l} \frac{[(A_1 - A_2)u_{rm} + 2Q_{eu}]t_d}{2} + \frac{Q_{eu} t_g}{2} \\ \frac{(A_1 u_{rm} + 2Q_{eu})t_d}{2} + \frac{Q_{eu} t_g}{2} \end{array} \right. \\ &= \left[\begin{array}{ll} 2\psi_{A1}\psi_{ur}^2\psi_{sp}\left[(1-\beta_2)\beta_1 + \frac{2\beta_1\psi_{ur}\psi_{sp}}{\beta} + \frac{\psi_{ur}^2\psi_{sp}}{(1-\beta_2)\beta^2}\right]\frac{E_i}{p_i} & (the\ rear - control\ type) \\ 2\psi_{A1}\psi_{ur}^2\psi_{sp}\left[\beta_1 + \frac{2(1+\beta_2)\beta_1\psi_{ur}\psi_{sp}}{\beta} + \frac{(1+\beta_2)^2\psi_{ur}^2\psi_{sp}^2}{\beta_2^2}\right]\frac{E_i}{p_i} & (the\ dual - control\ type) \end{array} \right., \end{aligned} \tag{2.65}$$

$$\left[\Delta V_o = \left[\begin{array}{l} \frac{[(A_1 - A_2)u_{im} - Q_{eu}](t_i - t_g)}{2} \\ \frac{(A_1 u_{im} - Q_{eu})(t_i - t_g)}{2} \end{array} \right. \right]$$

$$\begin{aligned} \Delta V_o &= \left[\begin{array}{l} \frac{[(A_1 - A_2)u_{im} - Q_{eu}](t_i - t_g)}{2} \\ \frac{(A_1 u_{im} - Q_{eu})(t_i - t_g)}{2} \end{array} \right. \\ &= \left[\begin{array}{ll} 2(1-\beta_2)\psi_{A1}\psi_{sp}\left[1 - \frac{\psi_{ur}^2\psi_{sp}}{(1-\beta_2)\beta}\right]^2 \frac{E_i}{p_i} & (the\ rear - control\ type) \\ 2\psi_{A1}\psi_{sp}\left[1 - \frac{(1+\beta_2)\psi_{ur}^2\psi_{sp}}{\beta}\right]^2 \frac{Ei}{Pi} & (the\ dual - control\ type) \end{array} \right.. \end{aligned} \tag{2.66}$$

When

$$A_2 u_{rm} \le Q_{eu}, \tag{2.67}$$

that is, when the maximum flow rate required for the piston's return acceleration is less than the supply flow rate of the pump (after deducting the leakage flow rate), it can be observed from Fig. 2.1 (Q_{eu} is in the dashed position) that there is no oil discharge process in the accumulator during the return stroke. In this case, the value obtained for ΔV_{ro} represents the charging volume rather than the discharging volume. Here, the sum of ΔV_{ri}, ΔV_{ro} and ΔV_i corresponds to the charging volume and is numerically equal to the discharging volume ΔV_o. For calculation, let $\Delta V_{ri} = \Delta V_{ro} = 0$ and $\Delta V_i = \Delta V_o$.

It is worth noting that under the ideal condition of $k_y = k_o = 0$ and no leakage, the aforementioned calculations can be expressed as:

$$\Delta V_{ri} = \frac{\psi_{sp}}{\beta^2} \cdot \frac{E_i}{p_i}, \tag{2.68}$$

$$\Delta V_{ro} = (\frac{\psi_{sp}}{\beta} - \psi_{ur})^2 \frac{E_i}{p_i} \tag{2.69}$$

$$\Delta V_i = \left(\psi_{sp} + \psi_{ur}\right)^2 \frac{E_i}{P_i}, \tag{2.70}$$

$$\Delta V_o = (1 - \psi_{sp})^2 \frac{E_i}{p_i}. \tag{2.71}$$

The calculation formulas for the charging and discharging oil volumes of the accumulator within one impact cycle have been derived above. It is evident that the discharging volume ΔV_o during the impact stroke phase or the charging volume ΔV_i during the return stroke phase is the largest. The larger of the two is taken as the design displacement ΔV_a, which serves as the basis for calculating the inflation volume. Assuming the ratio between the highest and lowest pressures during the operation of the accumulator is denoted as λ, and the adiabatic index of the gas is denoted as k (for nitrogen gas, $k = 1.4$), we can obtain

$$V_a = \frac{\lambda^{1/k} + 1}{\lambda^{1/k} - 1} \cdot \Delta V_a, \tag{2.72}$$

$$p_a = \frac{p_i}{2^k}. \tag{2.73}$$

The smaller the value of λ is, the smaller the fluctuations in working pressure, but it may result in an unrealistic large inflation volume. Typically, λ ranges from 1.2 to 1.35.

The series of formulas obtained in Sects. 2.2 to 2.5 of this chapter are primarily composed of dimensionless coefficients such as acceleration ratios and resistance coefficients, along with the impact energy, impact frequency (period), and working oil pressure of the hydraulic impact mechanism. This is advantageous for analysis and design.

2.6 Relationship Between Input Flow Rate and Other Parameters

When the input flow rate Q_i of a hydraulic impact mechanism varies, the working oil pressure p_i, the impact period (frequency), $T(f)$, and the impact energy E_i, also vary. The analytical relationships among these parameters will be derived in this section.

From Eqs. (2.13), (2.16) and (2.35), we obtain

$$u_{im}^2 = \frac{A_1 S_p P_i}{\psi_{A1}\psi_{sp} m_p}, \tag{2.74}$$

$$E_i = \frac{A_1 S_p P_i}{2\psi_{A1}\psi_{sp}}, \tag{2.75}$$

and

$$T = \sqrt{\frac{\psi_{A1} S_p m_p}{\psi_{sp} A_1 p_i}}. \tag{2.76}$$

With the substitution of Eqs. (2.75) and (2.76) into Eq. (2.57), we can obtain

$$Q_i = \varphi_l p_i + \varphi_p \sqrt{p_i}, \tag{2.77}$$

or

$$\varphi_l^2 p_i^2 - \left(\varphi_p^2 + 2\varphi_l Q_i\right) p_i + Q_i^2 = 0. \tag{2.78}$$

The solution is

$$p_i = \frac{\varphi_p^2 + 2\varphi_l Q_i - \sqrt{(\varphi_p^2 + 2\varphi_l Q_i)^2 - 4\varphi_l^2 Q_i^2}}{2\varphi_l^2}, \tag{2.79}$$

where

$$\varphi_p = \left[\begin{array}{ll} \frac{\psi_{ur}^2}{\beta}\sqrt{\frac{\psi_{sp} A_1^3 S_p}{\psi_{A1} m_p}} & (the\ rear - control\ type) \\ \frac{(1+\beta_2)^2 \psi_{ur}^2}{\beta}\sqrt{\frac{\psi_{sp} A_1^3 S_p}{\psi_{A1} m_p}} & (the\ dual - control\ type) \end{array}\right. . \tag{2.80}$$

With the above equations, we can determine parameters such as the impact energy, impact frequency (period), and working pressure p_i of a hydraulic impact mechanism with different input flow rates Q_i. Conversely, we can calculate the required input flow rate for a specified working pressure with Eq. (2.77).

When leakage is small, that is, $\varphi_l = 0$, Eq. (2.78) becomes a linear equation in terms of pressure p_i. The approximate solution is

$$p_i \approx \frac{Q_i^2}{\varphi_p^2}. \tag{2.81}$$

In fact, under the ideal condition without any losses, the exact relation between p_i and Q_i is

$$p_i = \left[\begin{array}{l} \frac{(1+\beta)^3 m_p}{2\psi_{sp}^2 A_1^3 S_p} Q_i^2 \ (the\ rear - control\ type) \\ \frac{m_p}{2\psi_{sp}^2 A_1^3 S_p} Q_i^2 \ (the\ dual - control\ type) \end{array} \right. . \tag{2.82}$$

In testing hydraulic impact mechanisms, experimental data have shown that the working oil pressure is approximately proportional to the square of the input flow rate. Equations (2.81) and (2.82) theoretically confirm this relationship.

So far, as the "three-step method" proposed in this book, a linear mathematical model has been established to describe the piston motion of hydraulic impact mechanisms. Next, further research will be conducted on several important issues of hydraulic impact mechanisms according to the fundamental equations mentioned above.

Chapter 3
Application of the Three-Step Method in Analysis and Design of Hydraulic Impact Mechanisms

3.1 Energy Loss and Efficiency

Expressions for several types of energy losses and efficiencies of the hydraulic impact mechanism will be further derived in this section, and the main factors affecting these losses and efficiency will be analyzed.

1. **Energy Consumption from Local Pressure Loss, E_c**

The internal oil passages of hydraulic impact mechanisms are complex, resulting in significant local pressure losses during fluid flow. In comparison, due to the relatively short oil passages, the pressure losses along the flow path can be neglected. However, the complexity of the flow channels and the drastic fluctuations in flow rate make precise calculation methods unsuitable for linear analysis methods, such as the three-step method. Since local losses are proportional to the square of the flow velocity, it is evident that the maximum velocities during the piston's return stroke and impact stroke, u_{rm} and u_{im} respectively, have the greatest influence on local pressure losses. Assuming the average cross-sectional flow area of the impact mechanism is A (determined by the nominal diameter of the design), and the total coefficient of local losses is ζ, the following simplified method can be used for calculation.

Taking the impact stroke phase of a dual-control hydraulic impact mechanism as an example, the local energy loss caused by a piston with velocity u within time t_i is

$$E = \int_0^{t_i} 0.5\zeta\rho\left(\frac{A_1u}{A}\right)^2 Q_i dt = 0.5\zeta\rho A_1^2\frac{Q_i}{A^2}\int_0^{t_i}(\omega_i t)^2 \mathrm{d}t$$

$$= 0.5\zeta\rho A_1^2\frac{Q_i}{A^2}\int_0^{t_i}\left(\frac{u_{im}t}{t_i}\right)^2 dt = \frac{\zeta\rho A_1^2u_{im}^2Q_it_i}{6A^2} = \frac{4\zeta\rho\psi_{A1}^2\psi_{sp}E_i^2Q_i}{3A^2p_i^2T} \tag{3.1}$$

With this method, the local energy loss for one cycle can be calculated as:

Q. He, *Research and Design of Hydraulic Impact Mechanisms*,
https://doi.org/10.1007/978-981-96-4864-1_3

$$E_c = \frac{\varphi_c E_i^2 Q_i}{p_i^2 T}, \tag{3.2}$$

where

$$\varphi_c = \left[\begin{array}{ll} \frac{4\zeta\rho\psi_{sp}\psi_{A1}[\beta_2^2\psi_{ur}^3+\beta(1-\beta_2)^2(1+\beta_1\psi_{ur}^3)]}{3\beta A^2} & (the\ rear-control\ type) \\ \frac{4\zeta\rho\psi_{sp}\psi_{A1}[\beta_2^2\psi_{ur}^3+\beta(1+\beta_1\psi_{ur}^3)]}{3\beta A^2} & (the\ dual-control\ type) \end{array} \right. \tag{3.3}$$

2. **Energy Consumption Due to Return Oil Resistance and Comprehensive Resistance, E_o and E_y**

The return oil resistance, which is reflected by the return oil resistance coefficient k_0 in Sect. 2.3, occurs in the chamber of the piston connected to the return oil. The energy consumption resulting from this can be expressed as:

$$E_o = \varphi_o E_i, \tag{3.4}$$

where

$$\varphi_o = \left[\begin{array}{ll} \frac{2k_0\psi_{ur}^2\psi_{A1}\psi_{sp}}{\beta} & (the\ rear-control\ type) \\ 2k_0(1+\beta_2-\beta_1\psi_{ur}^2)\psi_{A1}\psi_{sp} & (the\ dual-control\ type) \end{array} \right. \tag{3.5}$$

The resistance reflected by the comprehensive resistance coefficient k_y is mainly generated in the chamber of the piston connected to high-pressure oil, and the energy consumption resulting from this can be expressed as:

$$E_y = \varphi_y E_i, \tag{3.6}$$

where

$$\varphi_y = \left[\begin{array}{ll} 2k_y(1+2\beta_2+\beta_1\psi_{ur}^2)\psi_{A1}\psi_{sp} & (the\ rear-control\ type) \\ 2k_y[1+(\beta_1+\frac{\beta_2}{\beta})\psi_{ur}^2]\psi_{A1}\psi_{sp} & (the\ dual-control\ type) \end{array} \right. . \tag{3.7}$$

3. **Energy Loss Due to Leakage, E_l**

According to Eq. (2.55), the energy consumption within one impact cycle resulting from this is

$$E_l = V_l p_i = \varphi_l p_i^2 T, \tag{3.8}$$

where

$$\varphi_l = \left[\begin{array}{ll} k_{2l}+2k_{1l}\psi_{sp}(1+\beta_1\psi_{ur}) & (the\ rear-control\ type) \\ 2\psi_{sp}\left[\frac{k_{2l}\psi_{ur}}{\beta}+k_{1l}(1+\beta_1\psi_{ur})\right] & (the\ dual-control\ type) \end{array} \right. . \tag{3.9}$$

4. **Energy Loss Due to Stroke Loss, E_u**

The energy consumed by it within one impact cycle is

$$E_u = V_u p_i = \varphi_u E_i. \tag{3.10}$$

where

$$\varphi_u = \begin{bmatrix} 2\psi_{A1}\psi_{sp}(\beta_2 - \beta_1\psi_{ur}^2) \ (the\ rear - control\ type) \\ 2\psi_{A1}\psi_{sp}\psi_{ur}^2(\frac{\beta_2}{\beta} - \beta_1) \ (the\ dual - control\ type) \end{bmatrix} \tag{3.11}$$

5. **Volumetric Efficiency η_V, Mechanical Efficiency η_m, Pressure Efficiency η_p, and Overall Efficiency η**

By analyzing the primary types of energy losses during one cycle, the volumetric efficiency, mechanical efficiency, pressure efficiency, and overall efficiency of the hydraulic impact mechanism can be determined.

$$\eta_V = \frac{V_e}{V_e + V_u + V_l} = \begin{bmatrix} \frac{(A_1-A_2)S_p}{A_1S_r+\varphi_l p_i T} \\ \frac{A_1S_p}{(A_1+A_2)S_r+\varphi_l p_i T} \end{bmatrix}$$
$$= \begin{bmatrix} \frac{2\beta(1-\beta_2)\psi_{A1}\psi_{sp}E_i}{2\psi_{A1}\psi_{sp}\psi_{ur}^2E_i+\varphi_l\beta p_i^2T} & (the\ rear - control\ type) \\ \frac{2\beta\psi_{A1}\psi_{sp}E_i}{2(1+\beta_2)\psi_{A1}\psi_{sp}\psi_{ur}^2E_i+\varphi_l\beta p_i^2T} & (the\ dual - control\ type) \end{bmatrix}. \tag{3.12}$$

For the sake of analysis convenience, the input oil pressure p_i used earlier does not actually include the (average) pressure drop caused by local resistance losses Δp_c.

$$\Delta p_c = \frac{E_c}{Q_i T}. \tag{3.13}$$

The total hydraulic power input for one cycle is

$$E_{in} = E_i + E_c + E_l + E_y + E_o. \tag{3.14}$$

The reason why the total input energy does not include the energy loss due to stroke loss (E_u) is because the direct cause of stroke loss energy (E_u) is the comprehensive resistance coefficient k_y (which will be seen later as the main factor reducing the return stroke distance). This portion of energy consumption is already calculated in E_y.

Considering the energy consumption E_o, E_y due to return oil resistance and comprehensive resistance, the mechanical efficiency is given by

$$\eta_m = 1 - \frac{E_o + E_y}{E_{in}}. \tag{3.15}$$

Considering the energy consumption E_c due to local resistance losses, the pressure efficiency is given by

$$\eta_p = 1 - \frac{E_c}{E_{in}}. \tag{3.16}$$

It should be noted that the overall efficiency of the hydraulic impact mechanism cannot be calculated by multiplying the three efficiencies mentioned above. This is because the energy loss caused by the coefficient k_y should not be double-counted. Therefore, when calculating the overall efficiency, the energy loss due to stroke loss E_u must be deducted. Then the overall efficiency is

$$\eta = \frac{E_i}{E_{in}} = \frac{E_i}{E_i + E_c + E_l + E_y + E_o}. \tag{3.17}$$

Using Eq. (2.57) along with the expressions for the comprehensive resistance loss coefficient (φ_y), return oil resistance loss coefficient (φ_o), leakage loss coefficient (φ_l), and local resistance loss coefficient (φ_c), another expression for the overall efficiency can be obtained,

$$\eta = \frac{E_i T^2 p_i^3}{\varphi_l T^3 p_i^5 + (1 + \varphi_y + \varphi_o) E_i T^2 p_i^3 + \varphi_l \varphi_c E_i^2 T p_i^2 + \varphi_c \varphi_q E_i^3}. \tag{3.18}$$

From these analyses and derivations, it can be concluded that for a given impact energy and impact frequency (period), the expression for overall efficiency of a hydraulic impact mechanism is actually a function of two important design variables: the return-impact acceleration ratio (β) and the working oil pressure (p_i).

3.2 Efficiency Analysis

The level of energy conversion efficiency is clearly an important indicator for assessing the performance of hydraulic impact mechanisms. The energy loss in four forms of hydraulic impact mechanisms is analyzed and summarized in the previous section. In this section, various graphs will be established by the aforementioned equations to analyze the influence of several important parameters on energy loss and energy utilization efficiency in hydraulic impact mechanisms.

The results in the graphs are calculated based on a hydraulic impact mechanism with an impact energy of 500 J and an impact frequency of 30 Hz. In each analysis below, except for the parameter being discussed, the other parameters of the impact mechanism remain constant.

1. Return-Impact Acceleration Ratio β

In the aforementioned analysis, the dimensionless parameter β, the ratio of the return acceleration to the impact acceleration, is frequently mentioned and undoubtedly has a significant impact on various performances of the impact mechanism. Figures 3.1 and 3.2 were plotted with the aforementioned equations on a computer. These diagrams visually display how various energy losses and efficiencies of the "rear-control" and "dual-control" hydraulic impact mechanisms change with the value of β. The solid line in the figures represents the "rear-control" and the dashed line represents the "dual-control" (the same applies to subsequent figures).

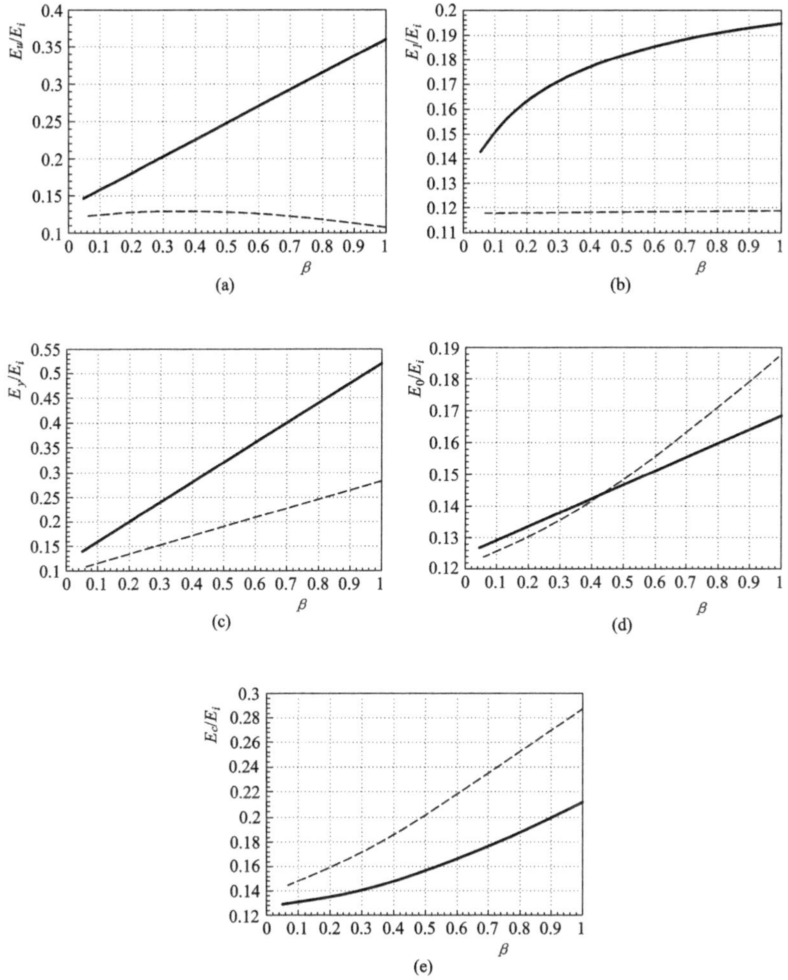

Fig. 3.1 Relationship between various energy losses and β

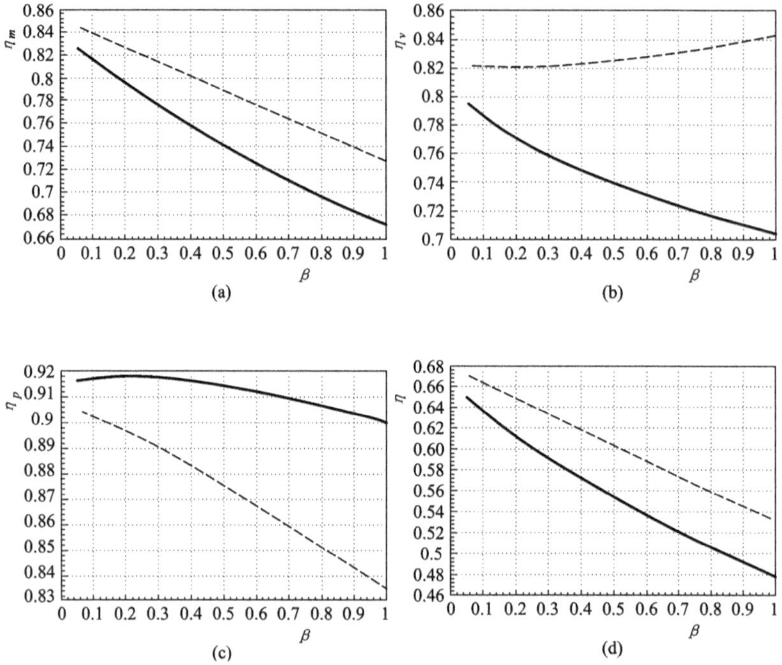

Fig. 3.2 Relationship between various efficiencies and β

In each figure, the horizontal axis represents the return-impact acceleration ratio β ranging from 0 to 1, and the vertical axis represents the ratio of various energy losses to the output impact energy, such as $\frac{E_u}{E_i}$. Among them, the stroke loss energy E_u and leakage loss energy E_l of the "dual-control" type hydraulic impact mechanism do not vary significantly with the value of β. However, other losses increase as the β value increases, with the magnitude of increase differing between the "rear-control" and "dual-control" types.

Compared to the "dual-control" type, the "rear-control" type exhibits higher magnitudes of energy losses, except for the local resistance loss energy E_c, as the value of β increases. Among the various efficiencies, the "rear-control" type has higher pressure efficiency than the "dual-control" type, while its volumetric efficiency, mechanical efficiency, and overall efficiency are lower. Moreover, as the β value increases, the difference between the two types becomes slightly larger.

2. **Oil Pressure p_i**

Figures 3.3 and 3.4 respectively illustrate the variations of energy losses and efficiencies with increasing supply oil pressure. The maximum supply oil pressure p_i for calculations is 21 MPa, and the values of β are all taken at 0.38. The horizontal axis represents the ratio of the supply oil pressure to the maximum supply oil pressure, denoted as $\frac{p_i}{p_{imax}}$.

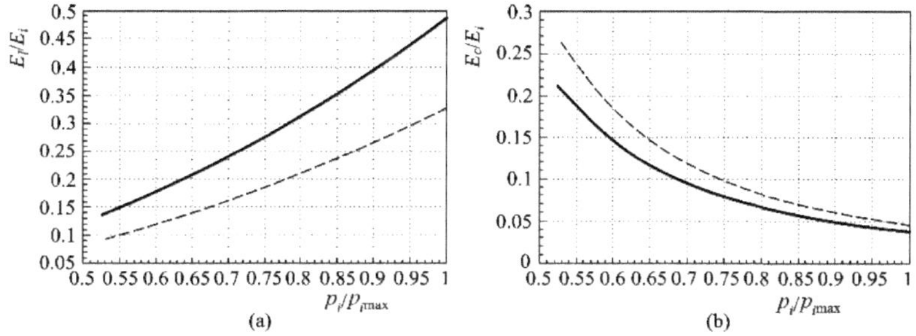

Fig. 3.3 Relationship between leakage loss, local resistance loss and oil pressure

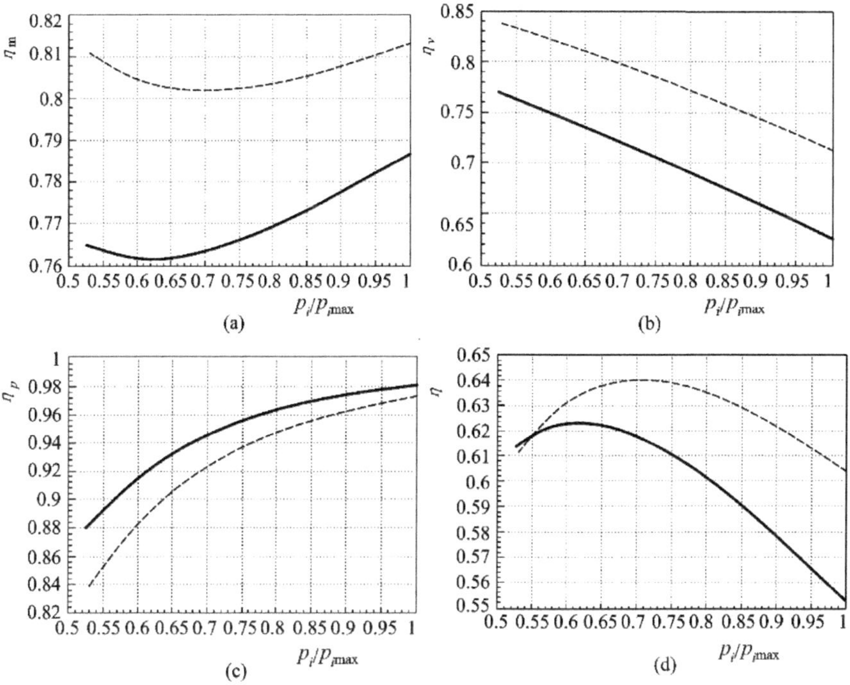

Fig. 3.4 Relationship between various efficiencies and oil pressure

The calculation expressions for stroke loss energy E_u, comprehensive resistance loss energy E_y, and return oil resistance loss energy E_o indicate that they are independent of the oil pressure p_i, so there is no need to represent them in charts.

Figure 3.3 shows that leakage loss increases with increasing pressure, while local resistance loss decreases with increasing pressure.

Figure 3.4 reflects that as the pressure increases, the volumetric efficiency decreases while the pressure efficiency increases. The mechanical efficiency curve

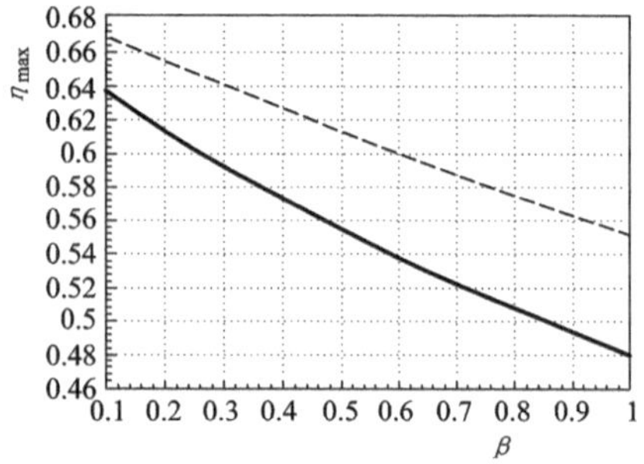

Fig. 3.5 Relationship between the maximum efficiency and β at a certain pressure

exhibits a valley, while the overall efficiency shows a peak. However, this peak does not vary significantly within a relatively wide range of pressure changes.

3. **Acceleration Ratio β, Pressure Range, and Maximum Efficiency**

Figure 3.5 represents the curves between different β values (horizontal axis) and the maximum efficiency (vertical axis) at a certain pressure. As the β value increases, this maximum efficiency decreases.

Figure 3.4d illustrates that the overall efficiency of the impact mechanism has a peak relative to the working pressure, but it is relatively flat. Therefore, instead of pursuing a single highest value, it is more practical to determine a pressure range that allows the overall efficiency to vary within a small range.

To further illustrate these points, Table 3.1 lists the maximum efficiency values corresponding to different β values and the pressure ranges in which they occur. The data in parentheses represent the calculated values for the 'dual-control' type.

The high-efficiency range in the table refers to the range of pressure p_i in which the efficiency value η_{max} is allowed to vary by 1.0%. This range is represented as the relative value $\frac{p_i}{p_{imax}}$.

4. **Combined Effect of β and p_i on Efficiency**

After analyzing the individual effects of β and p_i on the efficiency of the hydraulic impact mechanism, a further analysis is conducted to examine their combined effect on efficiency.

Figures 3.6, 3.7, 3.8, 3.9, 3.10 and 3.11, drawn based on Eq. (3.18), provide a visual and intuitive representation of the relationship between energy utilization efficiency and the two design variables—the return-impact acceleration ratio β and working pressure p_i—for rear-chamber-control and dual-chamber-control hydraulic impact mechanisms. In the figure, "D-C" represents the rear-control type, and "R-C" represents the dual-control type.The 3D plot in Fig. 3.6 uses β and p_i/p_{imax} as the x–y plane, with the mesh surface illustrating the variation in total efficiency η. Figures 3.7, 3.8, 3.9, 3.10 and 3.11 further refine the distribution of iso-efficiency curves on the x–y plane from the 3D plot in Fig. 3.6.

Figure 3.6a (rear-control type) and b (dual-control type) depict the relationship between the overall efficiency η and p_i, as well as β when the above four kinds of energy loss are relatively large. Figure 3.6c (rear-control type) and d (dual-control type) depict the occasions where all four kinds of energy loss are relatively small.

Table 3.1 Pressure Range for η_{max} at Different β Values

β	η_{max}	Range of high efficiency
0.1	0.64	0.500–0.690
	(0.67)	(0.525–0.762)
0.2	0.61	0.500–0.690
	(0.65)	(0.525–0.762)
0.3	0.59	0.500–0.690
	(0.64)	(0.550–0.786)
0.4	0.57	0.500–0.690
	(0.63)	(0.550–0.786)
0.5	0.55	0.500–0.714
	(0.61)	(0.550–0.809)
0.6	0.54	0.500–0.714
	(0.60)	(0.575–0.833)
0.7	0.52	0.524–0.714
	(0.59)	(0.575–0.833)
0.8	0.51	0.524–0.738
	(0.58)	(0.575–0.857)
0.9	0.49	0.524–0.738
	(0.56)	(0.575–0.857)
1.0	0.48	0.524–0.738
	(0.55)	(0.600–0.881)

These figures show that the overall efficiency of the impact mechanism has an extremum. To provide a clearer view, the iso-efficiency curves of the total efficiency surfaces in each figure are projected onto the $\beta - p_i/p_{imax}$ plane and correspondingly presented in the four subplots of Fig. 3.7.

From the figures, it can be observed that the magnitude of energy losses not only affects the level of efficiency but also shifts the distribution of high-efficiency regions towards larger values of β and p_i.

Since the magnitude of total energy consumption can alter the shape of the iso-efficiency curves, how does the variation in each of the four types of energy consumption individually affect the shape of the iso-efficiency curves? Figures 3.8, 3.9, 3.10 and 3.11 respectively depict the distribution of iso-efficiency curves when the comprehensive resistance coefficient k_y, return oil resistance coefficient k_0, local resistance coefficient ζ, and leakage coefficient k_l is small, while the other coefficient values are large. For example, Fig. 3.8 shows the distribution of iso-efficiency curves when the comprehensive resistance coefficient k_y is very small, while the return oil resistance coefficient k_0, local resistance coefficient ζ, and leakage coefficient k_l are large. Comparing them with Fig. 3.7, it can be observed that k_y and k_0 have almost no effect on the distribution of the iso-efficiency curves, while ζ has the greatest influence. It causes the efficiency surface to become ridge-like without any peak, forming

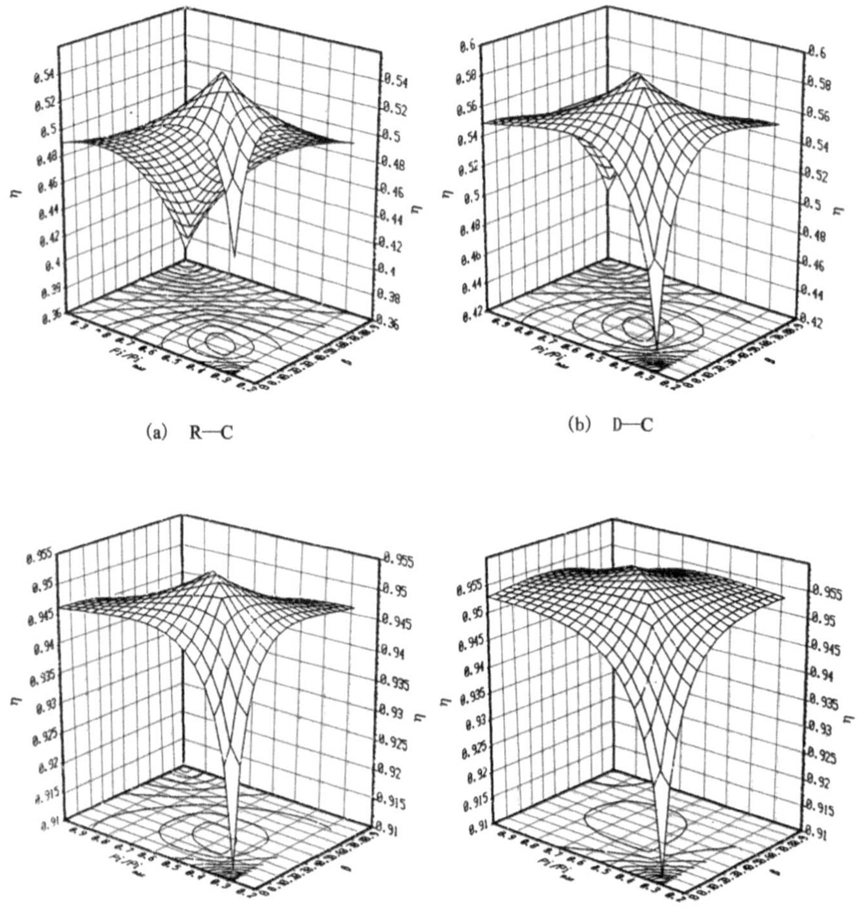

(a) R—C (b) D—C

(c) R—C (d) D—C

Fig. 3.6 Relationship between the overall efficiency η and the oil pressure p_i, acceleration ratio β (in a three-dimensional plot)

a 'ridge' of iso-efficiency. The coefficient k_l significantly shifts the high-efficiency region towards larger values of β and p_i.

In summary, when various energy losses are at a general level, the hydraulic impact mechanism exhibits the following characteristics in the relationship between overall efficiency η and return-impact acceleration β, oil pressure p_i:

① There is a high-efficiency peak region; if β or p_i is too small or too large, the efficiency η will decrease.

② The high efficiency region of the rear-control type hydraulic impact mechanism is within the range of $\beta = 0.15 \sim 0.50$ and $\frac{p_i}{p_{imax}} = 0.38 \sim 0.64$, while the high efficiency region of the dual-control type is within the range of $\beta = 0.24 \sim 0.76$ and $\frac{p_i}{p_{imax}} = 0.45 \sim 0.82$.

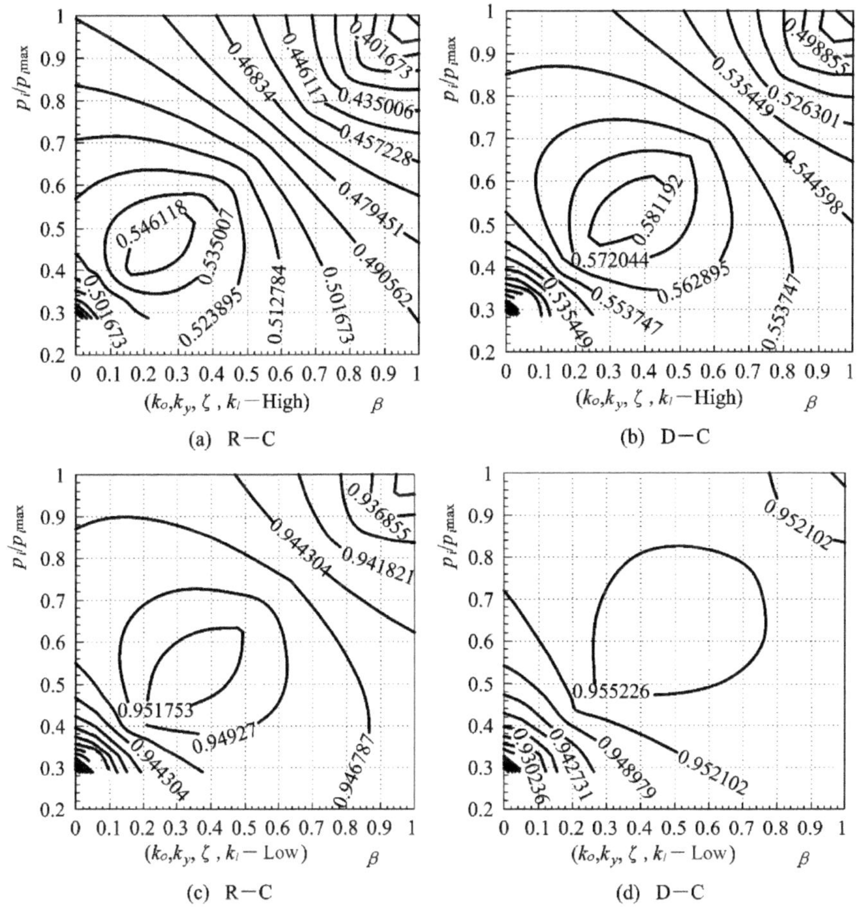

Fig. 3.7 Relationship between the overall efficiency η and oil pressure p_i, and acceleration ratio β in terms of various high or low losses (represented by the distribution of equivalent efficiency curves)

③ Under the same conditions, the efficiency of the rear-control type is lower than that of the dual-control type, and the density of iso-efficiency curves in the latter is lower than that in the former.

④ Even in the majority of regions far from the high-efficiency region, the decrease in efficiency is not very pronounced.

⑤ Due to the relatively wide high-efficiency region and its less sensitivity to changes in β and p_i, it provides designers with a significant range of choices in terms of efficiency criteria. In other words, pursuing a single set of designs that aim for the "highest" efficiency unnecessary and impractical.

5. **The Comprehensive Resistance Coefficient k_y and the Stroke Loss Energy E_u**

From the following analysis, it can be observed that the coefficient k_y has a significant effect on the performance of the impact mechanism. Figure 3.12a–c respectively

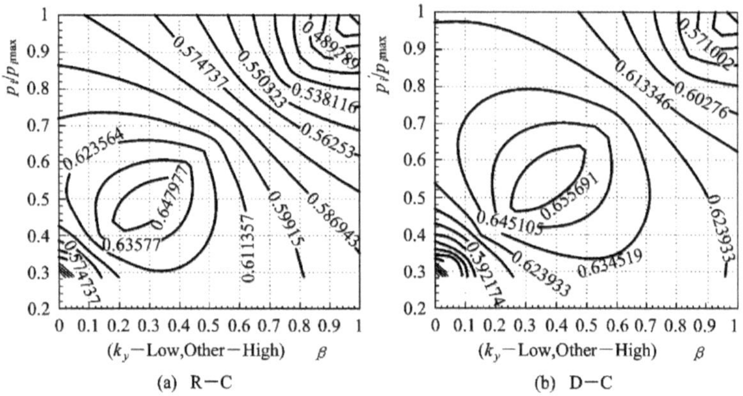

Fig. 3.8 Relationship between the overall efficiency η and oil pressure p_i, acceleration ratio β when the comprehensive resistance loss is low (represented by the distribution of equivalent efficiency curves)

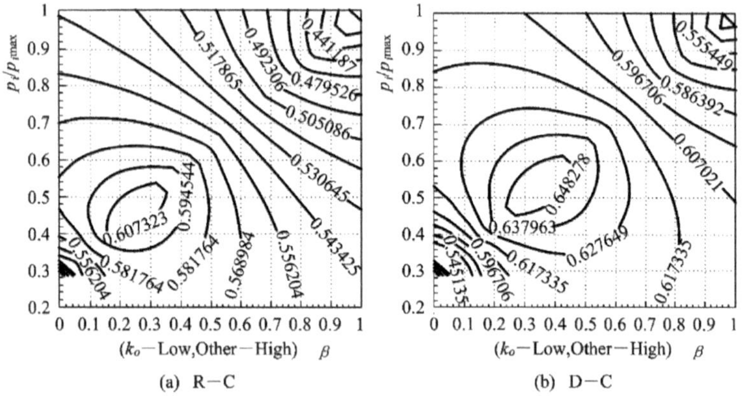

Fig. 3.9 Relationship between overall efficiency η and oil pressurep_i, the acceleration ratio β, when the return oil resistance loss is low (represented by equal efficiency distribution curves)

depict the variation of piston strokes for different values of k_y in the range of 0 to 0.12. The return acceleration stroke S_r and the total stroke S_p increase linearly as k_y increases, reaching a maximum of 1.04 and 1.16 times their values at $k_y = 0$, respectively. In contrast, the return braking stroke S_d decreases linearly, reaching a minimum of only 0.7 times its value at $k_y = 0$. These opposite trends indicate that the magnitude of the k_y coefficient greatly affects the recovery of energy generated at the return acceleration phase. Figure 3.12d directly shows the linear increase in stroke loss energy E_u due to the increase in the k_y coefficient, eventually exceeding twice its value at $k_y = 0$.

In addition, the return oil resistance, local pressure loss, and leakage have no or very little effect on the stroke loss energy with the analysis.

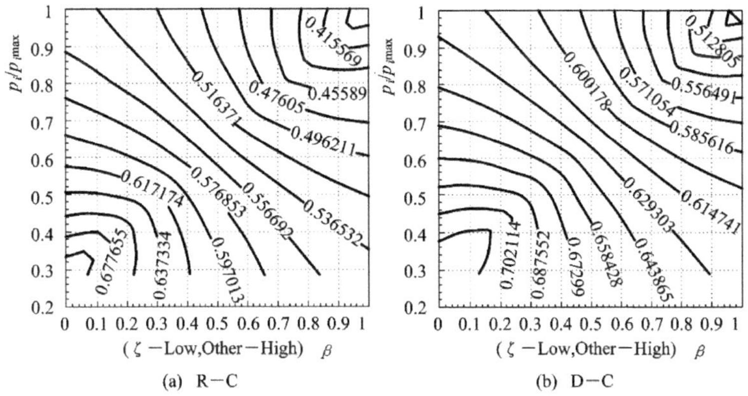

Fig. 3.10 Relationship between overall efficiency η and oil pressurep_i, the acceleration ratio β, when the local resistance loss is low (represented by equal efficiency distribution curves)

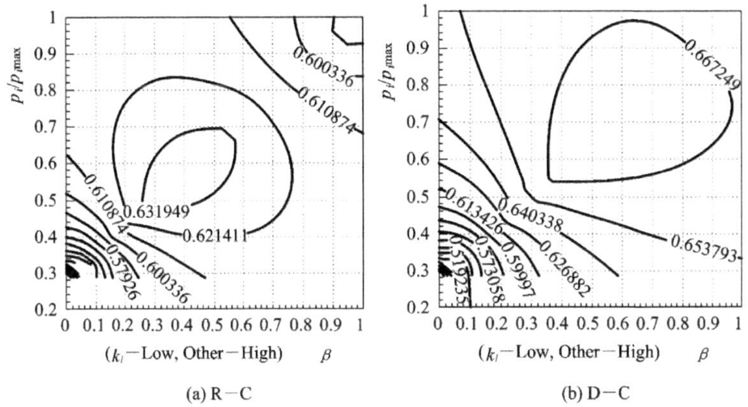

Fig. 3.11 Relationship between overall efficiency (η) and oil pressure p_i, acceleration ratio β when leakage losses are low (represented by equal efficiency distribution curves)

3.3 Accumulator Charging and Discharging Volumes and Acceleration Ratio β

The role of the accumulator is crucial to the performance of hydraulic impact mechanisms. It is a special type of diaphragm accumulator, with its diaphragm vibrating thousands of times per minute—up to nearly ten thousand times at maximum—and with a relatively large vibration amplitude. Consequently, various approaches have been explored to examine the relationship between the accumulator and relevant parameters of the impact mechanism in order to identify ways to reduce the frequency of oil charging and discharging as well as the volume of oil charged and discharged (thus reducing the amplitude). In this section, we will examine this issue from another perspective, aiming for a more detailed and practical analysis.

The basic parameters of the hydraulic impact mechanism analyzed below are the same as those in the previous chapter. Additionally, unless otherwise stated, the

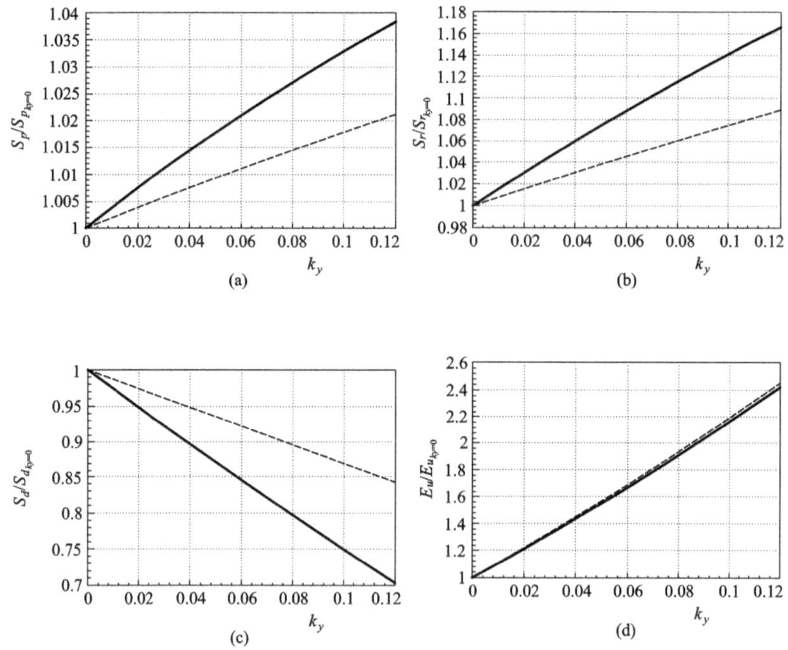

Fig. 3.12 Influence of the comprehensive resistance coefficient on the stroke and stroke loss

curves depicted by solid lines in the following graphs represent the rear-control type of impact mechanism, while the dashed lines represent the dual-control type.

1. **Charging and Discharging Times of the Accumulator Within One Impact Cycle and Its Characteristic Value β^***

Under normal conditions, the accumulator undergoes charging and discharging twice within one impact cycle. In other words, its diaphragm vibrates twice. However, as seen in the analysis of the accumulator in the previous chapter, if the design meets the condition of Eq. (2.67), the diaphragm of the accumulator vibrates only once within one impact cycle, completing one cycle of charging and discharging. Clearly, this is highly beneficial for extending the service life of the accumulator diaphragm.

The Eq. (2.67) can also be expressed as:

$$\left[\begin{array}{ll} \beta_2 \leq \frac{\psi_{ur}\psi_{sp}}{\beta} & (rear - control\ type) \\ \frac{\beta_2}{1+\beta_2} \leq \frac{\psi_{ur}\psi_{sp}}{\beta} & (dual - control\ type) \end{array}\right. . \tag{3.19}$$

In fact, it is the relationship between β and k_0, k_y. For given values of k_0 and k_y, Eq. (3.19) can be transformed into a quartic inequality in terms of β to determine the β value that satisfies Eq. (2.67) (which can be more conveniently solved with an iterative algorithm on a computer). This value is referred as the characteristic value of the accumulator's operating cycles, denoted as β^*.

The characteristic value β^* can be analytically obtained under the ideal condition of $k_0 = k_y = 0$ and no leakage. According to Eq. (2.51), regardless of rear-control or dual-control, Eq. (3.19) can be expressed uniformly as follows:

$$\frac{\beta}{1+\beta} - \frac{\psi_{ur}\psi_{sp}}{\beta} \leq 0. \tag{3.20}$$

Solving this inequality yields a characteristic value of β^*:

$$\beta^* \leq \frac{1}{3}. \tag{3.21}$$

For practical hydraulic impact mechanisms, 1/3 represents the maximum value of the characteristic value β^* that the accumulator charges and discharges oil only once respectively within one impact cycle. As k_0 and k_y increase, the values of β satisfying Eq. (3.19) will be decreased.

2. **The Minimum Oil Charging and Discharging Volume of the Accumulator and the Characteristic Value $\boldsymbol{\beta}^{**}$**

Firstly, analyzing Fig. 3.13a, b, they respectively depict the relationship between the accumulator charging volume ΔV_i (represented by solid lines), discharging volume ΔV_o (represented by dashed lines), and the return-impact acceleration ratio β for the rear-control type and dual-control type impact mechanisms. The vertical axis is constructed by the ratio of ΔV_i and ΔV_o to their calculated values at a small β value ($\beta = 0.05$).

The overlapping segment of the curves in the figure (with values of β less than 1/3) indicates the working region where the accumulator only charges and discharges once within one impact cycle, and the horizontal coordinate at the end of the overlapping lines represents the characteristic value β^*. In this segment, the charging and discharging volumes decrease synchronously as β increases. After the β^* point, the discharging volume ΔV_o continues to decrease more gradually, while the charging volume ΔV_i sharply rises again after a sudden drop and intersects with the discharging curve. Since the accumulator's design displacement is determined by the larger of the required ΔV_o and ΔV_i, it is evident that the accumulator's design displacement is minimized at the intersection point. The return-impact acceleration ratio β corresponding to the intersection point becomes another special value known as the accumulator displacement characteristic value, denoted as β^{**}.

Clearly, the triangular regions in Figs. 3.13a, b cannot be considered as the design range for the accumulator displacement. However, Fig. 3.13c depicts the relationship curve between the designed displacement ΔV_a of the accumulator and the acceleration ratio β after excluding the triangular region.

What factors are related to the magnitude of the characteristic value β^{**} of the accumulator displacement? Through computational analysis, it is found to be related to the magnitude of the comprehensive resistance coefficient k_y and the return oil resistance coefficientk_0. Figure 3.14 illustrates the individual impact of k_y and k_0

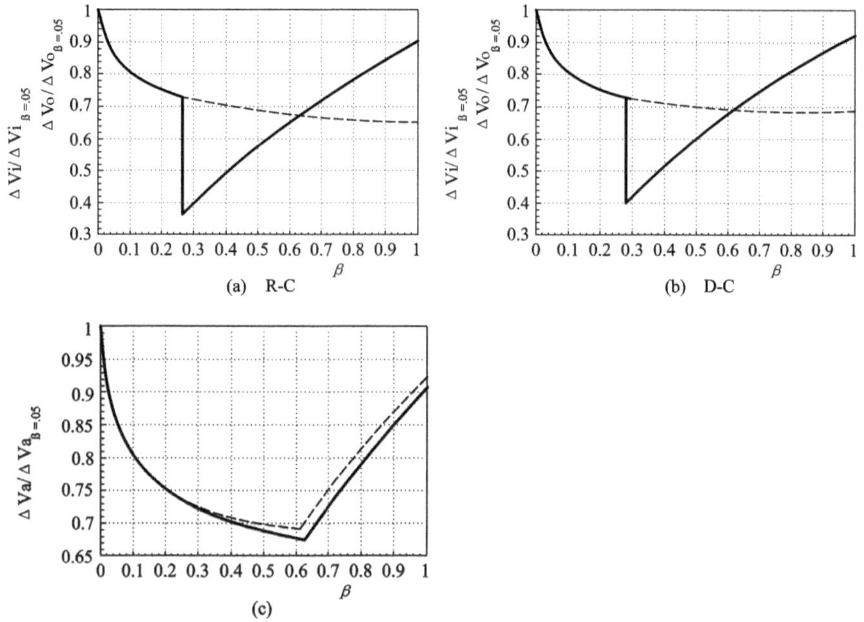

Fig. 3.13 Relationship between the charging and discharging volumes of the accumulator and β

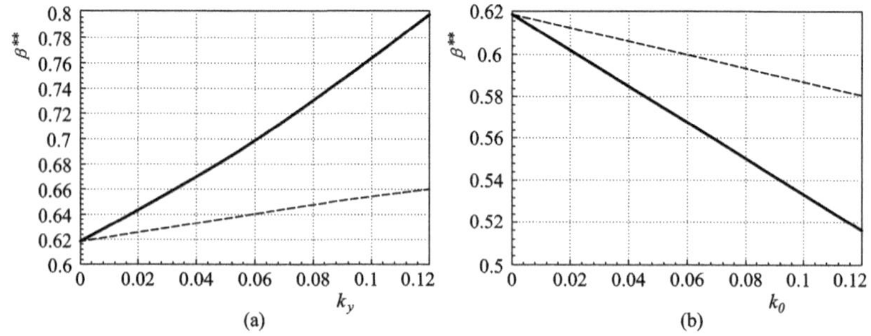

Fig. 3.14 The individual effects of comprehensive resistance and return oil resistance on the characteristic value of accumulator displacement

onβ^{**}, with k_y having a more pronounced effect, particularly on the rear-control type of impact mechanism. Despite the tendency of k_0 to decreaseβ^{**}, its overall effect, combined withk_y, leads to an increase in β^{**} with increasingk_y. Figure 3.15 presents a contour plot depicting the combined influence of k_y and k_0 onβ^{**}. It shows that β^{**} increases with the rise of both resistance coefficients, with a minimal effect observed for the dual-control type.

Clearly, determining the characteristic value β^{**} involves finding the abscissa β of the intersection point between the charging curve ΔV_i and the discharging curve ΔV_i

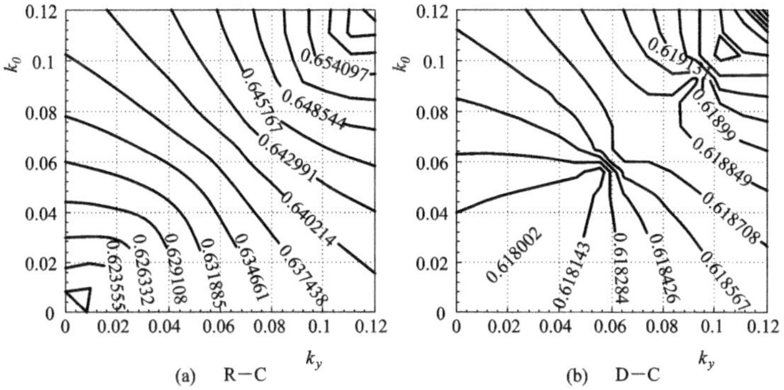

Fig. 3.15 The combined effects of comprehensive resistance and return oil resistance on the characteristic value of accumulator displacement

of the accumulator. Equating the right-hand side of Eq. (2.65) with the right-hand side of Eq. (2.66) yields the following equation:

$$\begin{bmatrix} 2(1+\beta_1\psi_{ur})\psi_{ur}^2\psi_{sp} - \beta(1-\beta_2)(1-\beta_1\psi_{ur}^2) = 0\ (rear - control) \\ 2(1+\beta_2)(1+\beta_1\psi_{ur})\psi_{ur}^2\psi_{sp} - \beta(1-\beta_1\psi_{ur}^2) = 0\ (dual - control) \end{bmatrix}. \tag{3.22}$$

Equation (3.22) can be transformed into a higher-order equation in terms of the acceleration ratio β, where the other coefficients are composed of the two resistance coefficients, k_y and k_O. In other words, the characteristic value β^{**} is determined by these two coefficients, and it can be numerically solved on a computer.

It can be observed that the value of β^{**} when the two resistance coefficients are zero lies between 0.6 and 0.62 as shown in Fig. 3.14. In fact, based on Eqs. (2.70) and (2.71), the intersection equation for the accumulator charging and discharging curves under the ideal condition of $k_O = k_y = 0$ can be obtained as

$$\left(\psi_{sp} + \psi_{ur}\right)^2 = \left(1 - \psi_{sp}\right)^2.$$

By substituting Eqs. (2.26) and (2.27) into the above equation and simplifying, we have

$$\beta^2 + \beta - 1 = 0. \tag{3.23}$$

At this point, the characteristic value of the accumulator displacement happens to be a somewhat magical number known as the golden ratio,

$$\beta^{**} = \frac{\sqrt{5}-1}{2} \approx 0.618. \tag{3.24}$$

However, in practical hydraulic impact mechanisms, the β^{**} value that minimizes the accumulator displacement is generally larger or slightly smaller than this number. It can be considered that 0.618 is the minimum value of the displacement characteristic for the accumulator.

3. **Selection of the Return-Impact Acceleration Ratio β**

We have theoretically studied the conditions for the accumulator of a hydraulic impact mechanism to undergo only one charging and discharging cycle per working cycle and to have the minimum designed displacement. Now, we will discuss how to select the acceleration ratio β with practical considerations.

The two characteristic values β^*, β^{**} corresponding to different resistance coefficients k_0 and k_y are listed in Table 3.2 (the data in the second row are calculated of "dual-control" type).

How to select the return-impact acceleration ratio β when designing hydraulic impact mechanisms? The data in Table 3.2 show that if we want to minimize the vibration frequency of the accumulator diaphragm, β^* must be less than 1/3, while if we want the accumulator displacement to be low, β^{**} must be greater than 0.618. Clearly, it is impossible to simultaneously satisfy both requirements. To address this issue, let's examine a calculation example. For a heavy-duty hydraulic impact mechanism with an impact energy of 500 J, the two characteristic values of the accumulator are $\beta^* = 0.299(0.312)$ and $\beta^{**} = 0.625(0.612)$ (the values in parentheses represent the values of the dual-control type). Within the interval ($\beta^{**} - \beta^*$), eight equally spaced β values are chosen, and the corresponding values of the return stroke discharge ΔV_{ro} and the designed displacement ΔV_a of the accumulator are calculated and listed in Table 3.3 (the data in the second row for each parameter represent the values of the dual control type).

The data in Table 3.3 indicate two characteristics of the accumulator operating within the region between the two characteristic values:

- During the return acceleration phase, there is a discharge process, but the discharge volume ΔV_{ro} is small. Even in the middle of the region, the discharge volume is not significant. However, this discharge process must also include the pressure oil for driving the control valve shifting during the return stroke.
- As the β value decreases, the designed oil displacement ΔV_a increases, but the increase is minimal.

Table 3.2 Relationship between the characteristic value of β and the resistance coefficients of k_y and k_0

K_0	0.0	0.02	0.04	0.06	0.08	0.10	0.12
k_y	0.0	0.02	0.04	0.06	0.08	0.10	0.12
β^*	0.333	0.322	0.31	0.297	0.283	0.268	0.251
	0.333	0.326	0.319	0.312	0.303	0.294	0.283
β^{**}	0.618	0.624	0.631	0.638	0.646	0.655	0.665
	0.618	0.618	0.618	0.619	0.619	0.620	0.620

Table 3.3 Relationship between β and the discharge of the accumulator at the characteristic value interval

β	0.299	0.335	0.372	0.408	0.444	0.480	0.516	0.553	0.589	0.625
	0.312	0.346	0.379	0.412	0.446	0.479	0.512	0.545	0.579	0.612
ΔV_{ro} (cm^3)	0.15	0.54	1.09	1.75	2.49	3.28	4.11	4.96	5.82	6.70
	0.12	0.43	0.87	1.45	2.02	2.66	3.34	4.03	4.74	5.44
ΔV_a (cm^3)	31.8	31.4	31.1	30.9	30.6	30.4	30.2	30.0	29.9	29.7
	31.5	31.2	31.0	30.8	30.6	30.5	30.4	30.3	30.2	30.1

Generally, the hydraulic impact mechanism is equipped with a flat diaphragm-type accumulator. The charging and discharging volume ΔV can be expressed as the product of the (projected) area A_a of the diaphragm and the displacement Δh of the diaphragm relative to the middle position during charging and discharging. That is

$$\Delta V = A_a \Delta h.$$

In this example, the impact mechanism is equipped with two accumulators, each with a diaphragm area of $A_a = 95$ cm^2. The control valve requires less than 2 cm^3 of pressure oil for each reversing operation during the return stroke. Even when considering the extreme value of $\beta = 0.299$ (for the rear-control type) and adding the pressure oil required for valve reversal, the ΔV value amounts to only $31.8 - 29.7 + 2 = 4.1$ cm^3. In this case

$$\Delta h = \frac{4.1}{95 \times 2} = 0.0215\,(\text{cm}).$$

That is to say, the diaphragm of the accumulator hardly moves. Additionally, the design capacity of the accumulator when $\beta = 0.299$ is approximately 2cm^3 larger than the minimum design displacement when $= \beta^{**}0.625$, and the additional diaphragm displacement required to discharge oil is even smaller. If we also consider the flow compensation effect resulting from the volume changes of hydraulic fluid and rubber hoses mentioned later in this book, the amplitude of the diaphragm's oscillation becomes even more negligible.

Let's observe Fig. 3.13c again. When β is between β^* and β^{**}, the ΔV_a curve declines gradually, while outside of this range, it changes much more rapidly, especially increasing faster when $\beta > \beta^{**}$.

It can be seen that the β value between the two characteristic values has no substantial impact on the charging and discharging frequency of the accumulator or its design displacement. Therefore, when selecting β in the design, it is unnecessary to adhere strictly to the principle of charging and discharging only once per cycle or minimizing the design capacity, especially since both objectives cannot be achieved simultaneously.

Taking into account the high-efficiency range discussed earlier, selecting $\beta = 0.36 \sim 0.55$ is a more suitable range. If β is chosen too small, there will be a significant difference in the area between the front and rear chambers of the piston, resulting in a large difference in step diameter on the piston. This is detrimental to the transmission of impact energy and the strength of the piston. If it is chosen too large, the drawbacks of increased piston stroke and reduced efficiency will become prominent.

It should be noted that, where installation space permits, increasing the A_a value of the accumulator is clearly beneficial for reducing the deformation of the diaphragm.

3.4 Application of "Three-Step Analysis" in the Design of Hydraulic Impact Mechanisms

1. Application Method of the "Three-Step Analysis"

Clearly, the series of calculation equations derived from the 'Three-Step Analysis' of the hydraulic impact mechanism are highly useful for the design of hydraulic impact mechanisms. The methods and steps for applying these equations in design will be introduced in this section.

(1) Determine the required impact energy E_i and impact frequency f for a hydraulic impact mechanism.
(2) Determine the rated input oil pressure, p_i (including local pressure losses), based on the performance of hydraulic components.
(3) Select the appropriate maximum impact velocity of the piston u_{im} (generally not exceeding 10 m/s and not lower than 6 m/s).
(4) Select k_0 and k_y within the range of return oil resistance coefficient $k_0 = 0.05 \sim 0.12$ and comprehensive resistance coefficient $k_y = 0.06 \sim 0.12$.
(5) Determine the maximum piston diameter and the length and clearance of the fit (sealing) in the front and rear bushings and cylinder liners, based on the diameter of the impact energy transmission body (such as the drill rod) and the structure of the impact mechanism. Use the leakage formula for an eccentric annular gap to calculate the leakage coefficients k_{2l} and k_{1l} of the front and rear chambers (with an eccentricity ratio of 0.8 to 0.95).
(6) Select the total local resistance loss coefficient ζ. Due to the complex and variable oil passages inside the hydraulic impact mechanism, determining the exact value of ζ is challenging. It is recommended to choose ζ within the range of 180 to 400 based on experience. Additionally, determine the nominal diameter of the oil passages and calculate the corresponding flow area.
(7) Calculate and determine the two characteristic values β^* and β^{**} of the accumulator. Use the trial method to solve Eqs. (3.19) and (3.22) to determine β^* and β^{**}. Then, select a value for β within the range of these two characteristic values. Based on Eqs. (2.37) and (2.40), or (2.46) and (2.49), calculate β_1 and β_2.

(8) Calculate the required parameters with the corresponding equations in following steps.

- Equation (2.1) → piston mass m_p;
- Equations (2.14) and (2.16) → stroke coefficient ψ_{sp} and stroke S_p;
- Equations (2.15) and (2.22) → return stroke velocity coefficient ψ_{ur} and maximum return stroke velocity u_{rm};
- Equations (2.28) and (2.29) → return stroke control oil port position S_c and impact stroke control oil port position S_{ic};
- Equation (2.35) or (2.44) → pressure area coefficient of the rear chamber ψ_{A1};
- Equation (2.36) or (2.45) → pressure area coefficient of the front chamber ψ_{A2};
- Equations (3.2) and (3.3) → total local lost energy E_c and its coefficients φ_c;
- Equations (3.4) and (3.5) → return oil resistance energy loss E_o and its coefficients φ_o;
- Equations (3.6) and (3.7) → comprehensive resistance energy loss E_y and its coefficient φ_y;
- Equations (3.8) and (3.9) → total leakage loss energy E_l and its coefficient φ_l;
- Equations (2.57) and (2.58) → initial calculation of input flow rate Q_i and flow coefficients rate φ_q;
- Equation (3.13) → the oil pressure difference Δp_c caused by local pressure loss. Subtract this oil pressure difference from the initially selected oil pressure p_i before proceeding with the following calculations.
- Equation (2.57) → final determination of input flow rate Q_i (excluding the oil consumption of the control valve, generally accounted for about 10% of Q_i; therefore, the whole machine input flow rate is approximately1.08–1.12 Q_i);
- Equations (2.65), (2.66), (2.72), and (2.73) → Design displacement of the accumulator ΔV_a, inflation volume V_a and inflation pressure p_a of the high-pressure accumulator;
- Equations (2.35) or (2.44) → oil pressure area of rear chamber A_1; Eq. (2.36) or (2.45) → oil pressure area of front chamber A_2;
- Based on A_1, A_2 and the maximum diameter d of the piston, the inner diameters d_1 and d_2 in the rear and front chamber bushings of the piston can be obtained and appropriately rounded.
- Equations (3.14), (3.17), or (3.18) → overall efficiency η (if the energy consumption of the control valve is considered, the result should be divided by 1.08 to 1.11);

Conversely, the β value of the existing hydraulic impact mechanism can also be estimated with Eqs. (2.38) and (2.47), and then the oil pressure, impact energy, impact frequency and other parameters can be calculated for different input oil flow rates with Eqs. (2.74) to (2.80).

The author compiled the above design methods into a computer program. The program list and instructions are shown in Appendix A at the end of this book.

Table 3.4 shows a comparison between a set of design values and test values for the YYG250 hydraulic impact mechanism.

The data in Table 3.4 indicate that the design method for the hydraulic impact mechanism derived from the "Three-Step Analysis" can be applied in practical engineering.

It is worth pointing out that the magnitude of the piston rebound after impact significantly affects the above calculation results. However, considering that the rebound of the hydraulic impact mechanism is minimal during actual work, such as in rock-breaking impacts, the coefficient of restitution, which measures the magnitude of piston rebound, has not been included in the above analysis.

2. **Summary of "Three-Step Analysis"**

The analyses in Chaps. 2 and 3 can be summarized in the following points:

(1) Dividing the piston motion cycle of the two common types of hydraulic impact mechanisms—'rear-chamber control' and 'dual-chamber control'—into three phases aligns more closely with actual working conditions. The series of mathematical equations based on this approach form a distinctive three-step analysis method with both theoretical and practical value.
(2) The dimensionless comprehensive parameter proposed by the author, the return-impact acceleration ratio β, makes the theoretical analysis of hydraulic impact mechanisms more abstract and universal. As a result, the design methods derived from this approach become more concise and practical.
(3) Studies on the relationship between stroke energy loss, volumetric efficiency, flow rate and pressure further reveal the motion characteristics of the hydraulic impact mechanism.
(4) The analysis of energy loss and efficiency for both types of mechanisms are instructive for optimizing the design scheme of the hydraulic impact mechanism.
(5) The calculation steps in the book can be relatively simple and accurate to complete the design of various parameters of the hydraulic impact mechanism.

Table 3.4 Comparison of Design and Test Values

	E_i /J	f /Hz	u_{im} $/(\mathrm{m \cdot s^{-1}})$	u_{rm} $/(\mathrm{m \cdot s^{-1}})$	p_i /Mpa	Q_i $/(\mathrm{L \cdot min^{-1}})$	A_1 $/\mathrm{cm}^2$	A_2 $/\mathrm{cm}^2$	η /%
Design Values	400	28.33	8.0	4.52	12.0	116	14.2	5.08	48.55
Test Values	396	28.71	7.82	4.32	12.1	120.8	14.83	5.06	47.58

Note The energy consumption of the control valve has been taken into account in the design flow rate and efficiency in the table, and the calculated values have been multiplied and divided by 1.1 respectively

Chapter 4
Analysis of Working Principle of Hydraulic Impact Mechanisms

4.1 Overview

1. **Assumptions for Establishing Mathematical Model of Impact Mechanisms**

A simulation model should strive to accurately reflect the actual physical processes of the simulated object. However, achieving a perfect representation is impossible, so some appropriate assumptions must be established. The assumptions are as follows:

(1) The viscosity of the hydraulic oil is unaffected by pressure, and the temperature is consistent throughout the machine's operation.
(2) Excluding the inlet hose, all other components are considered to be ideally rigid without deformation.
(3) The mass of the accumulator diaphragm is zero and there is no resistance to deformation.
(4) The mass of the oil is ignored in most cases and is treated as a lumped parameter only when it should be considered.
(5) The pump supplies a constant oil flow rate.
(6) The transmission time of the pressure wave in the oil is negligible.
(7) The impacted object is treated as a rigid body and the piston velocity drops to zero immediately after the impact.

2. **Basic Structure and Calculation Parameters Distribution Diagram of YYG250 Hydraulic Impact Mechanism**

Figure 4.1 illustrates the basic structure of the YYG250 impact mechanism. The upper part is the control valve, and the lower part is the piston, both having a constant-pressure chamber. The oil pressures in the constant-pressure chamber are denoted by p and p_2, while p_3 and p_1 represent the control chamber pressures of the valve and the piston, respectively.

The symbols in Fig. 4.1 are explained as follows:

Q. He, *Research and Design of Hydraulic Impact Mechanisms*,
https://doi.org/10.1007/978-981-96-4864-1_4

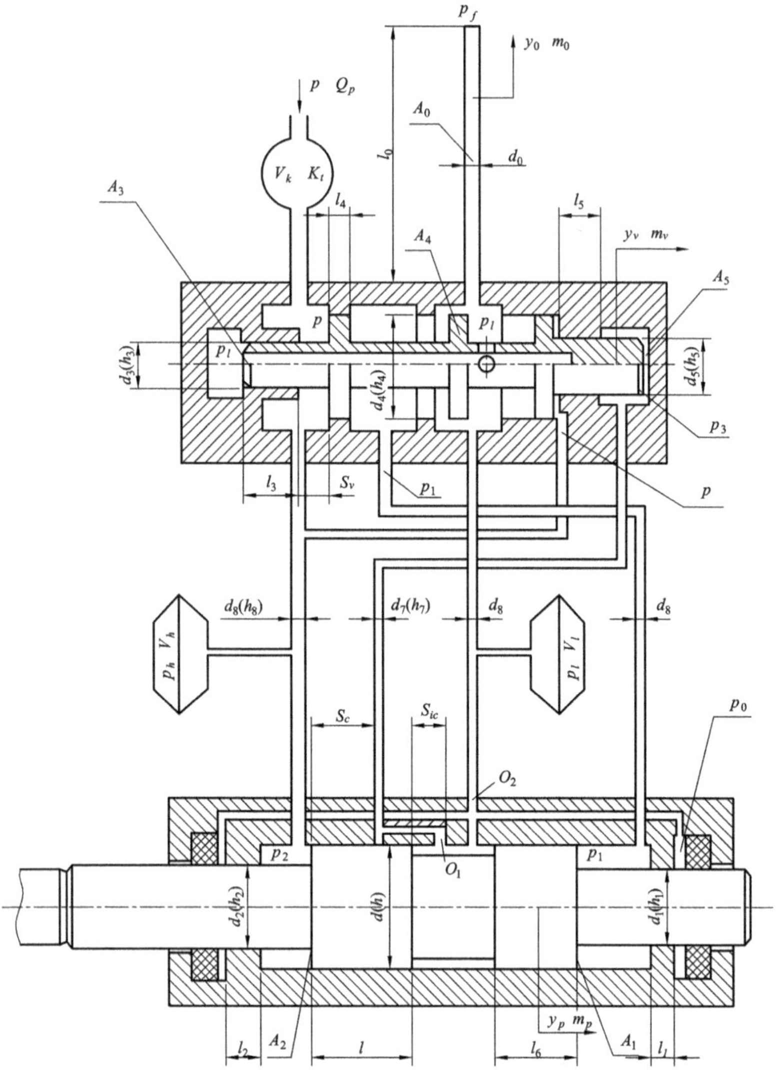

Fig. 4.1 Structure and parameter distribution of YYG250 hydraulic impact mechanism

- m_p, m_v, m_0: mass of the piston, valve spool, and the oil in the return pipe.
- y_p, y_v, y_0: displacement of the piston, valve spool, and the centroid of the oil in the return pipe.
- p_1, p_2, p_3: oil pressure in the chambers of the piston, valve.
- p: inlet pressure of the inlet hose, considered as input oil pressure of the impact mechanism, with pressure losses along the pipeline neglected.
- p_f: backpressure of the oil return filter.

- p_h, p_l: working pressure of the high-pressure and low- pressure accumulators, where p_l is also regarded as the oil return pressure;
- p_{ah}, p_{al}: inflation pressure of high-pressure and low-pressure accumulators.
- A_1, A_2: working areas of the rear and front chamber of the piston.
- A_3, A_4, A_5: areas of the three steps of the controls valve spool.
- A_0, d_0: cross-sectional area and inner diameter of the oil return pipe.
- $d, d_1, d_2, d_3, d_4, d_5$: diameters of each cylindrical sealing surface;
- $h, h_1, h_2, h_3, h_4, h_5$——radial clearance heights corresponding to each cylindrical sealing surfaces;
- $l, l_1, l_2, l_3, l_4, l_5, l_6$: sealing lengths for each cylindrical sealing surfaces;
- l_0: length of the oil return pipe;
- d_7, A_7: diameter and corresponding cross section area of the return stroke feedback oil passage.
- d_8, A_8: design diameter and corresponding area of the internal oil passage in the impact mechanism.
- Q_i: supply oil volume of the pump (considered as a constant value).
- S_r, S_d: acceleration and deceleration strokes of the piston during return stroke.
- S_p: stroke of the piston.
- S_v: stroke of the Spool.
- S_c: distance of the feedback hole from the control side of the piston during the return stroke (based on the piston being in the return start position).
- S_{ic}: distance of the feedback hole from the control side of the piston during the impact stroke (based on the piston being in the start position of return stroke).
- V_h, V_l: chamber volume of high-pressure and low-pressure accumulators.
- V_{ah}, V_{al}: inflation volume of the high-pressure accumulator and the low-pressure accumulator.
- V_k: volume of the inlet pipe and other high-pressure oil chambers (considered as a constant);
- K_e: volumetric elastic modulus of the oil.
- K_E: apparent volumetric elasticity coefficient of the inlet pipe and the oil within it.

4.2 Analysis of the Working State of Hydraulic Impact Mechanisms

What we commonly refer to as the return stroke and the impact stroke in hydraulic impact mechanisms are actually based on the motion phase of the impacting piston. The aforementioned three-step analysis method used in linear model analysis only divides the piston's motion into three phases: return stroke acceleration, return stroke deceleration, and impact stroke acceleration. However, the actual working process of a hydraulic impact mechanism involves the automatic control motion of three main moving bodies: the piston, valve, and accumulator. Therefore, its motion state within one impact cycle becomes much more complex. There are 12 distinct motion states from the beginning of the return stroke until the start of the next return stroke, with the striking end surface of the impacted object (shank adapter, drill rod, etc.) as the reference point. Taking the rear-control YYG250 hydraulic impact mechanism as an example, the characteristics of these 12 states are briefly described below (refer to Figs. 4.1 and 4.2).

State A: Valve is stationary; piston accelerates during return stroke.

At this time, the chamber A_5 of the control valve is connected to the return oil, meaning $p_3 = p_l$, under the force

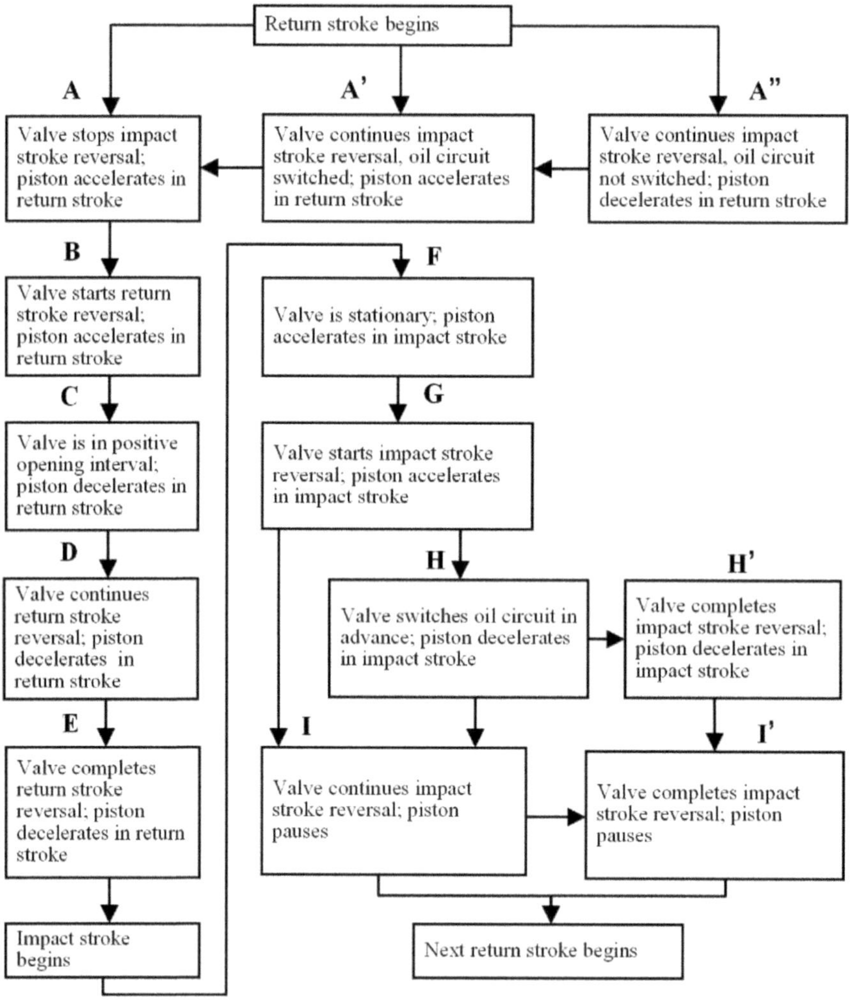

Fig. 4.2 YYG250 hydraulic impact mechanism working status

$$\begin{aligned} F_v &= p(A_4 - A_3) + p_l A_3 - p(A_4 - A_5) - p_l A_5 \\ &= (A_5 - A_3)(p - p_l) \end{aligned} \tag{4.1}$$

the valve is fully open in the right position, and the rear chamber of the piston is connected to the return oil, so $p_1 \approx p_l$.

As the piston velocity increases, the flow rate of oil discharged from its rear chamber also increases, and the value of p_1 rises due to the throttling effect of the valve port. However, under the hydraulic force.

$$F_p = p_2 A_2 - p_1 A_1 > 0, \tag{4.2}$$

the piston overcomes the viscous friction and other resistance, accelerates to the right.

State B: Valve begins impact stroke reversal; piston continues accelerating in return stroke.

When the piston accelerates through the return control stroke S_c, its return control surface A_2 opens the valve's reversing control oil passage d_7. Pressure oil feeds back from the piston's front chamber to the control chamber A_5, causing the valve spool to begin reversing to the left under the hydraulic force

$$\begin{aligned} F_v &= p_3 A_5 + p(A_4 - A_5) - p(A_4 - A_3) - p_l A_3 \\ &= A_3(p - p_l) + A_5(p_3 - p). \end{aligned} \tag{4.3}$$

Before the spool switches the oil circuit—meaning, before the spool enters the positive opening range z_0 (completing half of the travel $(S_v - z_0)/2$)—the rear chamber of the piston remains connected to the return oil. However, as the valve port gradually narrows, the discharge pressure p_1 of the piston's rear chamber gradually rises due to throttling effects at the orifice, and the difference between p_1 and the return pressure p_l increases progressively. Consequently, the value in Eq. (4.2) approaches zero. During this period, the piston continues its return stroke motion to the right. For convenience, we can define the valve displacement $y_v \leq (S_v - z_0)/2/2$ as the return acceleration phase. However, in fact, due to the variation of the valve port, the condition $F_p = p_2 A_2 - p_1 A_1 \leq 0$ actually occurs before $y_v = (S_v - z_0)/2$. In other words, the piston's return acceleration phase actually ends before this point, and it begins the return stroke deceleration motion with the maximum speed attained during the return stroke acceleration as the initial velocity.

State C: Valve enters its positive opening range; piston decelerates during return stroke.

This is a relatively short but distinctive working state, characterized by the connection of the front chamber, rear chamber, and return oil chamber. The pressure p_1 in the rear chamber and the flow rate variations between the chambers are complex, but $F_p < 0$

consistently holds. Further details can be found in the relevant sections discussed later.

State D: Valve continues reversing during return stroke; piston continues decelerating during return stroke.

At this moment, the valve spool has exited the positive opening range where the three chambers were connected. The rear chamber of the piston is isolated from the return oil chamber and is now connected to the front chamber. The pressure difference between p_1 and p_2 is no longer significant. The front and rear chambers of the piston are both connected to the pressure oil, forming a so-called differential connection. The force F_p expressed by Eq. (4.2) is now $F_p < 0$, indicating that F_p has become a noticeable hindrance to the piston's motion.

State E: Valve is stationary; piston continues decelerating during return stroke.

The control valve spool has completed its travel S_v, ending the return stroke reversal motion. The valve port is fully open, and the piston's velocity further decreases under the opposing force F_p until it reaches zero. This completes the return stroke motion, which consists of the return acceleration stroke S_r and the return deceleration (or braking) stroke S_d, preparing for the next impact. The distance traveled by the piston is referred to as the piston stroke S_p.

State F: Valve is stationary; piston accelerates during the impact stroke.

From the perspective of hydraulic circuitry, similar to the return braking phase, the two chambers of the piston remain in a differential connection during this phase. However, the piston begins to accelerate in the opposite direction with a much greater acceleration compared to the return acceleration phase. Furthermore, the direction of hydraulic force is opposite to that of resistance, unlike in the return braking phase, where the hydraulic force direction is the same as the resistance direction.

State G: Valve starts impact stroke reversal; piston continues to accelerate during the impact stroke.

The piston continues accelerating, and when the piston displacement $y_p = S_p - S_{ic}$, the impact stroke control surface on the piston opens the oil passage O_1, which connects to the valve's control oil passage d_7. Through the groove in the middle of the piston, O_1 also connects to the oil passage O_2, which is always connected to the return oil. The oil previously sealed in the valve spool chamber A_5 now connects to the return oil passage, resulting in $p_3 = p_l$. Under the force F_v represented by Eq. (4.1), the valve spool begins moving to the right, initiating impact stroke reversal. We aim for the displacement of the valve spool not to exceed half of its total travel when the piston completes the final stroke S_{ic} and impacts the target. Otherwise, the flow rate entering the rear chamber of the piston will be strongly throttled, resulting in $F_p = p_2A_2 - p_1A_1 > 0$, causing impact stroke deceleration.

The force on the piston in this state is $F_p = p_2A_2 - p_1A_1 < 0$.

State I: Valve continues impact stroke reversal; piston pauses.

The valve spool continues impact stroke reversal under the force determined by Eq. (4.1). The piston impacts the target, and the impact stress wave initially propagates from the impact end face of the piston towards its tail in the form of a compression wave. After reaching the tail, it propagates from the tail towards the impact end in the form of a tensile wave. Only after reaching the impact end does the piston detach from the end face of the target in an impact rebound form. It is evident that during the time when the impact stress wave travels back and forth along the entire length of the piston, the piston remains in a stationary state. The duration of the pause t_s can be calculated using Eq. (4.4).

$$t_s = \frac{2l_p}{c}, \tag{4.4}$$

where l_p is the length of the piston, and c is the propagation speed of the stress wave, typically taken as 5000 m/s in steel.

Under normal conditions, the distance traveled by the valve during the piston's stationary period should exceed half of its total stroke. This indicates that the oil circuit has been switched to allow the piston's rear chamber to flow back to the return oil, thereby enabling the piston to rebound smoothly.

Because the impacted body is assumed to be an ideally rigid body, the piston has no displacement. In reality, there is some displacement, extending the pause time by approximately 0.2 to 0.3 ms. For most hydraulic rock drill impact mechanisms, the impact pause time t_s is about 0.5 to 0.6 ms.

The issue of how to handle the spool's impact stroke reversal will be discussed later.

State I': Valve completes impact stroke reversal; piston pauses.

This is a state in which the piston and valve are paused.

State A': Valve continues impact stroke reversal but has switched the oil circuit; piston accelerates during return stroke.

This is another form of the beginning of the return stroke.

The piston completes its impact pause and begins to rebound, while the spool continues the impact stroke reversal motion, but the oil circuit has been switched. This is still the normal working state, with the force on the spool still determined by Eq. (4.1), and the force on the piston the same as in the previous state *A*.

The ten working states analyzed above—*A*, *A*′, *B*, *C*, *D*, *E*, *F*, *G*, *I*, and *I'* — all belong to the normal operating conditions of the hydraulic impact mechanism. However, if the design parameters are not appropriate, the following three abnormal working states may occur.

State A″: Valve continues the impact stroke reversal but has not yet switched the oil circuit; this is an abnormal working state at the beginning of the return stroke. State A″ is not included in the 12 states.

After the piston completes its impact pause, in addition to the two possible states of *A* and *A'* to start the return stroke, there is also an abnormal state, that is, the spool has not switched the oil circuit due to slow reversing, and the pressure p_1 in the rear chamber is still in a high-pressure state. At this time, the piston will decelerate with the rebound speed as the initial speed, and its speed will decrease to zero. If the valve still does not switch the oil circuit, the piston will have the so-called secondary impact. Of course, it can also be assumed that the valve switches the oil circuit during the secondary impact process, and the piston slows down because the force on the front chamber is greater than that on the rear chamber, or it can complete the secondary impact and entering a secondary pause….

State H: Valve switches the oil circuit in advance during the impact stroke reversal; piston decelerates.

This is an abnormal phenomenon in the impact stroke, a special case of the state *G*. We hope that the piston can achieve the maximum impact speed before impacting the impacted body, but if the spool reverses too early, the pressure of the rear chamber, p_1, will be reduced, and the piston will enter a deceleration state, which will affect the increase of the impact speed. This is certainly a working state that we do not want.

State H′: Valve completes the impact stroke reversal; piston decelerates during impact stroke.

It is another special case of state H, in which the valve stops moving.

4.3 Analysis of Changes in Flow and Pressure of Hydraulic Impact Mechanisms

As previously mentioned, one of the key characteristics of the hydraulic impact mechanism is the significant variation in internal oil flow rate. Next, we will analyze the changes in flow rate and pressure under several primary operating conditions.

The research results indicate that, for in-depth analysis of the motion principles of the hydraulic impact mechanism, the compressibility of the oil in the high-pressure chamber of the impact mechanism cannot be ignored. Therefore, it is necessary to fully consider the characteristic of expansion under pressure of the high-pressure rubber hose that supplies pressure oil to the impact mechanism. However, current research treats the flow rate input into the impact mechanism as a constant value, which is not consistent with the actual situation, both from a theoretical analysis and subsequent experimental analysis.

People widely recognize that the accumulator is an indispensable component for hydraulic impact mechanisms. It plays a crucial role in recovering energy during the return stroke, providing peak flow rate, and greatly improving energy utilization. However, the flow rate variations caused by oil compression and hose expansion due to pressure changes are often overlooked. These flow rate variations are referred as compensating flow rate Q_e in this book. In fact, the oil volume generated by the compensating flow rate Q_e is significant relative to the oil discharge (consumption) of one working cycle of the impact mechanism. Therefore, when establishing the mathematical model of the impact mechanism's motion, it is necessary to consider the variations in this compensating flow rate. In Chap. 7 of this book, a detailed discussion will be provided on the motion principle of the hydraulic impact system after the removal of the high-pressure accumulator. It can be observed that this compensating flow rate Q_e is significant and cannot be ignored.

To gain an in-depth understanding of the motion principle of the hydraulic impact mechanism, we will first qualitatively analyze the flow rate and pressure variations in several of its main operating states.

The flow rate and pressure variation trends during one operating cycle of the hydraulic impact mechanism are depicted in Fig. 4.3. Most of the symbols in the figure have the same meanings as in previous chapters. For convenience, they are explained again as follows: T (working cycle), t_r (return acceleration time), t_d (return deceleration time), t_i (impact stroke acceleration time), t_s (impact pause time), Q_i (pump supply flow rate), p (inlet pressure of the impact mechanism), t_{rg} and t_g (the moment when the flow rate required for piston motion in the return stroke and impact stroke phase equals the pump supply flow rate). *A*, *B*, *D*, and so on represent the positions of the various working states described in the previous section on the timing diagram.

The curves in Fig. 4.3a, b depict the case where the pump flow rate Q_i is less than the maximum flow rate required by the piston during the return stroke acceleration phase, while the curves in Fig. 4.3c, d illustrate the case where the pump flow rate is approximately equal to the maximum flow rate required by the piston during return stroke acceleration.

The trends in flow rate and pressure variations, as observed in Fig. 4.3, are described as follows:

In the *A* and *B* states during return stroke acceleration, when the pump supply flow rate Q_i is greater than the flow rate required for piston movement, the excess oil is pushed into the high-pressure accumulator C_h for energy storage, resulting in an increase in both the accumulator pressure p_h and the supply pressure p. It is worth noting that the increase in pressure causes expansion of the inlet hose diameter as well as compression of the oil volume in the high-pressure chamber, including the inlet pipe, which also accumulates a portion of high-pressure oil.

As the piston's velocity increases, after surpassing the moment t_{rg} in Fig. 4.3, the pump supply flow rate Q_i may become insufficient to meet the flow rate required for piston movement. The two portions of accumulated high-pressure oil mentioned above will then be released to compensate. Naturally, the pressure p_h and p will decrease. However, as long as the design parameters are chosen appropriately, as shown in the lower part of Fig. 4.3, by making the pump supply flow rate roughly equal to the maximum oil demand of the piston during the return stroke acceleration phase, the high-pressure accumulator will only have one charging process without significant discharge during the return stroke acceleration phase. As for the larger

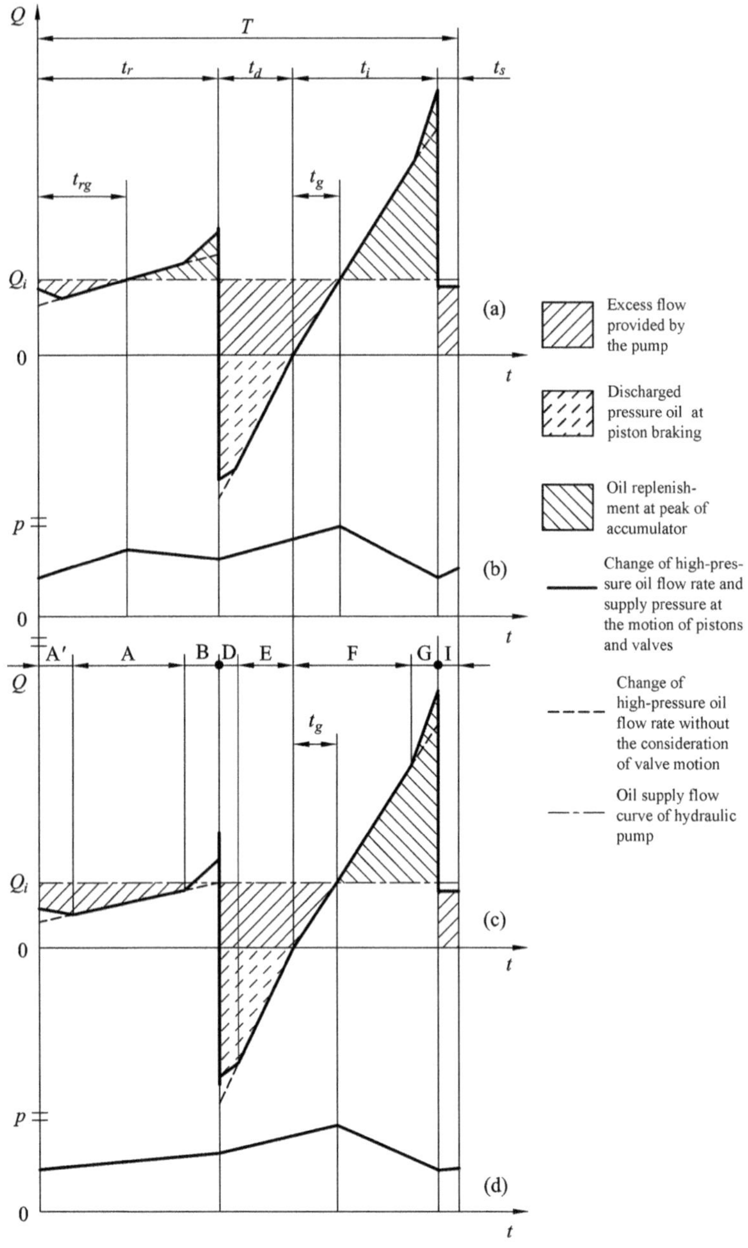

Fig. 4.3 Flow rate and pressure variation curves

peak flow rate required during the reversal of the control valve, due to the extremely short duration (only a few milliseconds) and the compensation of the flow rate caused by the expansion of oil in the high-pressure chamber and the contraction of the rubber hose, the high-pressure accumulator has almost no noticeable discharge process.

When the piston enters the *C, D*, and *E* states, it relies on the kinetic energy gained during the return stroke acceleration phase to overcome oil pressure resistance in the rear chamber, thereby decelerating. During this process, aside from leakage, the piston movement not only requires no external oil supply but also expels the oil from the rear chamber (minus the portion flowing to the front chamber through the differential connection). At this stage, the impact mechanism essentially acts as a pump, working together with the supply pump to press nearly all of these two portions of oil into the accumulator, causing the pressure to rise and eventually bringing the piston speed down to zero.

When the piston enters the impact stroke acceleration state, *F*, its velocity gradually increases from zero. Before reaching the moment t_g in Fig. 4.3, the pump's supply flow rate is still greater than the flow rate required by the piston. The excess oil continues to be pressed into the accumulator. At the time t_g, both the supply pressure p and the accumulator pressure p_h reach their maximum values, and the amount of high-pressure oil accumulated due to accumulator compression, oil compression, and hose expansion also reaches its peak. From time t_g to the end of impact stroke in state G, the piston velocity continues to increase, and the pump supply flow rate becomes much less than the demand. As the pressure decreases, the stored high-pressure oil is released to supplement the flow. At the moment of impact, the pressure drops to its lowest point, but the peak flow rate at this time is often several times greater than the pump's supply flow rate.

After the impact ends and the system enters state I, with the piston in a pause, nearly all of the pump's supply flow rate is again directed into the accumulator, except for the small amount needed for the valve spool reversal, causing a slight increase in pressure.

From the above analysis, it can be observed that within a short working cycle (as brief as a few milliseconds), the flow rate required for piston movement drops to zero twice and surges to a high level twice, with an even greater peak flow rate occurring at the end of the impact, particularly in hydraulic impact systems used as rock crushers. These flow rate variations are also accompanied by changes in pressure. As will be discussed later, the installation of a high-pressure accumulator significantly smooths out these pressure fluctuations.

4.4 Fundamental Equations of Motion

Referring to Fig. 4.1, the basic differential equations describing the combined motion of the three main moving bodies in the impact mechanism can be established based on the dynamic equilibrium equations of the moving bodies, the flow balance equation (i.e., the principle of liquid continuity), and the gas state equation:

Dynamic balance equation of the piston:

$$m_p \frac{d^2 y_p}{dt^2} + F_{sp}\left(\frac{dy_p}{dt}, \Delta p\right) + F_{lp}(p_l, p_1, p_2) = F_p(p_1, p_2). \tag{4.5}$$

High-pressure oil flow balance equation:

$$Q_i + Q_{ha}\left(\frac{dV_h}{dt}\right) - Q_l\left(\frac{dy_p}{dt}, \Delta p\right) = Q_p\left(\frac{dy_p}{dt}\right) + Q_v\left(\frac{dy_v}{dt}\right) + Q_e\left(\frac{dp}{dt}\right). \tag{4.6}$$

Gas state equation of the high-pressure accumulator:

$$p_h V_h^k = p_{ah} V_{ah}^k = C_h \tag{4.7}$$

Dynamic balance equation of the control valve:

$$m_v \frac{d^2 y_v}{dt^2} + F_{sv}\left(\frac{dy_v}{dt}, \Delta p\right) + F_{lv}(p) = F_v(p, p_l, p_3) \tag{4.8}$$

Dynamic balance equation of the oil in the return pipe:

$$m_0 \frac{d^2 y_0}{dt^2} + F_{s0}\left(\frac{dy_0}{dt}, p_f\right) = F_0(p_l) \tag{4.9}$$

Balance equation of return flow rate:

$$Q_l\left(\frac{dy_p}{dt}, \Delta p\right) + Q_{pv0}\left(\frac{dy_p}{dt}, \frac{dy_v}{dt}\right) = Q_0\left(\frac{dy_0}{dt}\right) + Q_{la}\left(\frac{dV_l}{dt}\right) \tag{4.10}$$

Gas state equation of the return accumulator:

$$p_l V_l^k = p_{al} V_{al}^k = C_l \tag{4.11}$$

The explanations for the above equations are as follows:

$m_p \frac{d^2 y_p}{dt^2}$: inertia force of the piston;

$F_{sp}\left(\frac{dy_p}{dt}, \Delta p\right)$: viscous friction resistance of the piston (a function of the relative velocity between the piston and the sealing surface and the oil pressure difference).

$F_{lp}(p_l, p_1, p_2)$: hydraulic clamping force on the piston;

$F_p(p_1, p_2)$: axial hydraulic thrust on the piston;

Q_i: supply flow rate of the pump (considered constant);

$Q_l\left(\frac{dy_p}{dt}, \Delta p\right)$: total leakage flow rate (leakage flow from the sealing surfaces of the piston and valve spool).

$Q_e\left(\frac{dp}{dt}\right)$: flow rate variation due to hose expansion and oil compression;

$Q_p\left(\frac{dy_p}{dt}\right)$, $Q_v\left(\frac{dy_v}{dt}\right)$: flow rates into the high-pressure chamber of the piston and valve, i.e., the high-pressure oil flow rates required to drive the piston and valve;

$Q_{ha}\left(\frac{dV_h}{dt}\right)$: charging (or discharging) flow rate of a high-pressure accumulator, i.e., the rate of change of its gas chamber volume with respect to time;

k: polytropic index (approximated as an adiabatic process for nitrogen $k = 1.4$);

C_h: initial state constant of the high-pressure accumulator;

$m_v\frac{d^2y_v}{dt^2}$: inertia force of the valve spool (m_v includes the mass of the oil pushed by the valve);

$F_{sv}\left(\frac{dy_v}{dt}, \Delta p\right)$: viscous friction resistance of the valve spool;

$F_{lv}(p)$: hydraulic clamping resistance of the valve spool;

$F_v(p, p_l, p_3)$: axial hydraulic thrust on the valve;

$m_0\frac{d^2y_0}{dt^2}$: inertial force of the oil column in the return pipe;

$F_{s0}\left(\frac{dy_0}{dt}, p_f\right)$: viscous resistance of the oil column in the return pipe and resistance from the return filter;

$F_0(p_l)$: thrust on the oil at the inlet of the return pipe;

$Q_{pv0}\left(\frac{dy_p}{dt}, \frac{dy_v}{dt}\right)$: discharge flow rate from the piston and valve spool to the return pipe;

$Q_0\left(\frac{dy_0}{dt}\right)$: flow rate passing through the return line;

$Q_{la}\left(\frac{dV_l}{dt}\right)$: charging and discharging flow rate of the return line accumulator;

C_l: initial state constant of the return accumulator.

The seven equations above are the fundamental motion equations of describing the movement of the impactor. The variables in the equations have different computational expressions depending on the various motion states of the impactor. Various detailed calculations and analyses will be performed, followed by the development of a specific, phase-by-phase mathematical model. Unless otherwise noted, all calculations in this book use the International System of Units (SI): length: meters (m); mass: kilograms (kg); time: seconds (s); force: newtons (N); viscosity: square meters per second (m^2/s) or pascal-seconds (Pa·s); pressure: pascals (Pa) or megapascals (MPa); flow rate: cubic meters per second (m^3/s).

4.5 Leakage, Viscous Friction Resistance, Hydraulic Clamping Resistance

Fundamental formulas:

$$Q_l = \frac{\pi dh^3}{12\mu l}(1 + 1.5\varepsilon^2)\Delta p \pm \frac{\pi dh}{2}u, \quad \text{(a)}$$

$$F_s = \frac{\pi \mu ld}{\sqrt{1 - \varepsilon^2 h}}u \pm \frac{\pi dh}{2}\Delta p, \quad \text{(b)}$$

$$F_l = \tau ld\,\Delta p. \quad \text{(c)}$$

Formulas (a), (b), and (c) represent, respectively, the calculation formulas for leakage, viscous resistance, and hydraulic clamping resistance for eccentric ring sealing surfaces with relative motion.

Where,

d: diameter of the inner cylindrical sealing surface;

h: radial clearance of the concentric ring;

u: relative velocity between the inner and outer cylindrical surfaces;

μ: dynamic viscosity coefficient of the hydraulic oil;

Δp: pressure differential across the sealing surface ends;

l: overlap length of the inner and outer cylindrical surfaces;

ε: eccentricity ratio between the two cylindrical surfaces; where $\varepsilon = e/h_0$, and e is the absolute eccentricity, assuming that the eccentricity of the valve spool and the piston with their respective cylindrical surfaces remains constant throughout the motion;

τ: resistance coefficient, which is the product of the clamping coefficient α and the friction coefficient ($\tau = 0.02$–0.025).

The principle for selecting "+" or "−" is as follows: when the direction of relative motion aligns with the direction of pressure difference, formula (a) takes "+" and formula (b) takes "−"; if the directions are opposite, formula (a) takes "−" and formula (b) takes "+".

Next, specific calculation formulas for leakage and viscous resistance are provided (see Fig. 4.1). Let $u_p = \frac{dy_p}{dt}$, $u_v = \frac{dy_v}{dt}$. Define Q_{l1}, Q_{l2}, Q_{lv} as the leakage rates from the rear chamber of the piston, the front chamber of the piston, and the valve, respectively. Then the total leakage rate is given by

$$Q_l = Q_{l1} + Q_{l2} + Q_{lv}. \quad (4.12)$$

Leakage occurs at the following eight locations: [d, h, l], [d, h, l_6], [d_1, h_1, l_1], [d_2, h_2, l_2], [d_3, h_3, l_3], [d_5, h_5, l_5], and two points at [d_4, h_4, l_4]. Additionally, because the valve's movement time is much shorter compared to the dwell time at both ends, the additional leakage caused by the valve's movement speed is not considered in the leakage calculation (in fact, much of this additional leakage would cancel itself out). In addition to the aforementioned eight locations, the generation of viscous frictional resistance and hydraulic clamping force takes into account the variation in valve opening area. In the valve spool, there are three sealing cylindrical surfaces with diameter d_4, but only two are actually involved in the sealing sliding, and the overlapping length of the seal varies, with the maximum being l_4. For simplicity, we consider the viscous frictional resistance and hydraulic clamping force on the two [d_4, l_4]. Furthermore, the sealing length of the piston cylinder with diameter d varies during operation due to the influence of the valve's reversal control oil passage, but this variation is not considered here. Based on the fundamental formulas (a), (b), and (c), the following specific calculation formulas for leakage, viscous frictional resistance, and hydraulic clamping force of the piston and valve spool can be derived.

$$\begin{aligned} Q_{l1} &= \frac{\pi\left(1+1.5\varepsilon^2\right)}{12\mu}\left(\frac{dh^3}{l_6}+\frac{d_1h_1^3}{l_1}\right)(p_1-p_l)+sgn\left(u_p\right)\frac{\pi}{2}(d_1h_1-dh)u_p \\ &= R_1(p_1-p_l)+sgn\left(u_p\right)R_2u_p \end{aligned} \tag{4.13}$$

$$\begin{aligned} Q_{l2} &= \frac{\pi\left(1+1.5\varepsilon^2\right)}{12\mu}\left(\frac{dh^3}{l}+\frac{d_2h_2^3}{l_2}\right)(p_2-p_l)+sgn\left(u_p\right)\frac{\pi}{2}(dh-d_2h_2)u_p \\ &= R_3(p_2-p_l)+sgn\left(u_p\right)R_4u_p \end{aligned} \tag{4.14}$$

$$\begin{aligned} Q_{lv} &= \frac{\pi\left(1+1.5\varepsilon^2\right)}{12\mu}\left(\frac{d_3h_3^3}{l_3}+\frac{2d_4h_4^3}{l_4}+\frac{d_5h_5^3}{l_5}\right)(p-p_l) \\ &= R_5(p-p_l) \end{aligned} \tag{4.15}$$

$$\begin{aligned} F_{sp} &= \frac{\pi\mu}{\sqrt{1-\varepsilon^2}}\left(\frac{dl}{h}+\frac{dl_6}{h}+\frac{d_1l_1}{h_1}+\frac{d_2l_2}{h_2}\right)u_p \\ &\quad + sgn\left(u_p\right)\frac{\pi}{2}\left[(d_2h_2-dh)(p_2-p_l)+(dh-d_1h_1)(p_1-p_l)\right] \\ &= B_1u_p+sgn\left(u_p\right)\left[B_2(p_2-p_l)+B_3(p_1-p_l)\right] \end{aligned} \tag{4.16}$$

$$\begin{aligned} F_{sv} &= \frac{\pi\mu}{\sqrt{1-\varepsilon^2}}\left(\frac{d_3l_3}{h_3}+\frac{2d_4l_4}{h_4}+\frac{d_5l_5}{h_5}\right)u_v-\frac{\pi}{2}d_3h_3(p-p_l) \\ &= B_6u_v-B_7(p-p_l) \end{aligned} \tag{4.17}$$

$$\begin{aligned} F_{lp} &= \tau\left[(dl+d_2l_2)(p_2-p_l)+(dl_6+d_1l_1)(p-p_l)\right] \\ &= B_4(p_2-p_l)+B_5(p_1-p_l) \end{aligned} \tag{4.18}$$

$$\begin{aligned} F_{lv} &= \tau\big[(d_3 l_3 + 2d_4 l_4)(p - p_l) + (d_5 l_5)(p - p_3)\big] \\ &= B_8(p - p_l) + B_9(p - p_3) \end{aligned} \tag{4.19}$$

Formulas (4.13) to (4.19) can be used for the calculation of each operating state mentioned earlier. The sign function $Sgn\big(u_p\big)$ in the formula takes into account that the direction of relative motion velocity of the sealing surface and the direction of pressure difference are sometimes the same and sometimes opposite when the state is different.

Of course, when performing state-specific calculations, some parameters in the formulas may vary significantly, sometimes with considerable differences. For example, the pressures p_3 and p_1 in the control chambers of the valve and piston can sometimes be high-pressure, while at other times they may be equal to the return pressure p_l.

4.6 Oil Pressure Differential Caused by Local Resistance—Local Resistance Loss

There are various types of local resistances inside the machine, including typical forms such as bends, valve ports, contractions, and expansions, as well as some non-typical local resistances. Among these, the 90° bend is the most common form of local resistance. However, the most common form of local resistance is the 90° bend. For ease of calculation and analysis, a single 90° bend loss is taken as the basic calculation unit (denoted as K_0), and all other local resistances are then converted into an equivalent number of 90° bend local resistances. Considering that the distance between adjacent resistances is very short and far from the transition length required for flow stabilization, the value of K_0 is taken as 2 to 3 times the normal value.

For ease of analysis, a calculation model was established, as shown in Fig. 4.4, to represent the flow distribution and corresponding local obstructions within the YYG250 impact mechanism. The model is explained as follows:

The pressures p, p_l, p_1, p_2, p_3, etc., correspond to those in Fig. 4.1 and rresent the oil pressure in various chambers;

ζ_1, ζ_2, ζ_3, ζ_4, ζ_5,and ζ_6 denote the coefficients of localized resistance at various points.

$Q_0, Q_1, Q_2, Q_3, Q_4, Q_5$ represent the flow rates passing through the local resistances mentioned above. Among them, the calculation of flow rates Q_3 and Q_4 is relatively complex. For the sake of simplicity, leakage flow, valve reversal flow, and flow variations caused by hydraulic oil compression and expansion are not taken into account when calculating local losses.

The nominal diameter of the pipes not indicated in the figure is d_8 and the cross-sectional area is A_8.

The basic formula is as follows:

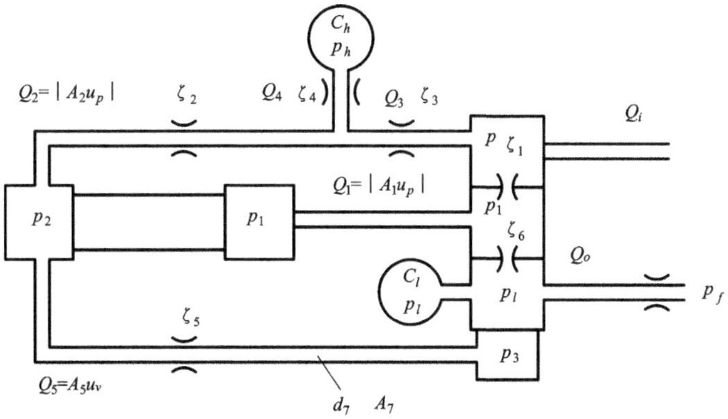

Fig. 4.4 Internal pressure difference calculation model of YYG250 hydraulic impact mechanism

$$\Delta p = \zeta \frac{\rho u^2}{2} = \zeta \frac{\rho}{2}\left(\frac{Q}{A}\right)^2,$$

where u: average fluid velocity;

ρ: density of liquids;

Q.: flow rate;

A: flow cross-sectional area;

ζ: local resistance coefficient.

Based on the established model and its basic formulas, the calculation formulas for pressure differentials can be derived in states.

It will be observed that not only do the magnitudes of the flow rates vary greatly in different locations, but the flow directions also frequently change, which brings considerable challenges to calculating the pressure difference. For the sake of analysis, when the flow direction is the same as the direction of the pressure difference, a positive value is taken, and when they are opposite, a negative value is taken.

1. **Return Acceleration Phase (Including States A, A’, and B)**

(1) Pressure Differential Between the Front Chamber (p_2) and the High-pressure Accumulator (p_h).

This pressure differential can be expressed as:

$$p_2 - p_h = \pm\zeta_2 \frac{\rho}{2}\left(\frac{Q_2}{A_8}\right)^2 \pm \zeta_4 \frac{\rho}{2}\left(\frac{Q_4}{A_8}\right)^2. \quad \text{(a)}$$

The selection of the signs on the right side of the equation (a) is determined by analyzing the flow direction of flow Q_2 and Q_4 (see Fig. 4.5).

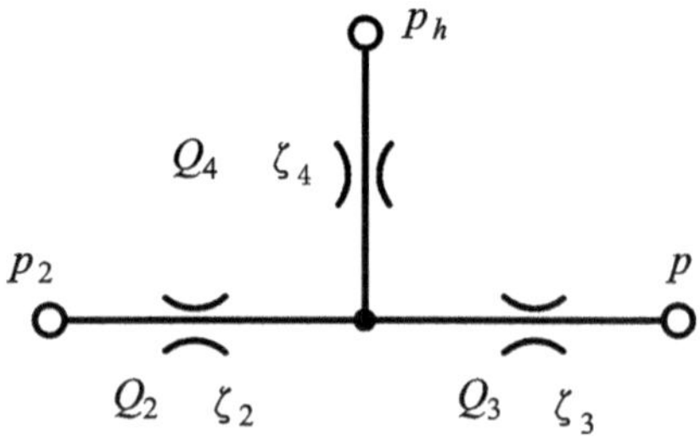

Fig. 4.5 Pressure differential calculation model between the valve inlet and the front chamber during the return acceleration phase

Since the flow Q_2 always flows to p_2 during the return acceleration phase, opposite to the direction of the pressure differential, a '−' sign is taken before the first term on the right of the equation.

The flow direction of Q_4 should be considered separately, as can be seen from $Q_4 = Q_i - Q_2$:

When $Q_4 \geq 0$, indicating that the pump supply flow rate exceeds the flow rate required by the piston movement in the front chamber, the excess part of flow Q_4 is directed into the accumulator, flowing in the same direction as the pressure differential. In this case, a '+'sign is applied to the second term on the right side of the equation.

When $Q_4 < 0$, indicating that the pump supply flow rate is no longer sufficient to meet the flow rate requirements of the piston's front chamber, the flow Q_4 generated by the accumulator's oil discharge to compensate flows in the opposite direction to the pressure differential, so the sign of the second term is '−'. By introducing the sign function, the above analysis can be uniformly represented by the Eq. (4.20).

$$\begin{aligned} p_2 - p_h &= sgn(Q_4)\zeta_4 \frac{\rho}{2}\left(\frac{Q_4}{A_8}\right)^2 - \zeta_2 \frac{\rho}{2}\left(\frac{Q_2}{A_8}\right)^2 \\ &= sgn\big(Q_i - A_2 u_p\big) K_4 (Q_i - A_2 u_P)^2 - K_2 (A_2 u_p)^2 \end{aligned} \tag{4.20}$$

(2) Pressure Differential Between the Valve inlet (p) and the Front Chamber (p_2).

The flow rate Q_3 in Fig. 4.5 represents the pump supply flow Q_i, and the flow direction of Q_2 and Q_3 is always consistent with the direction of the pressure differential. Therefore, we have

$$\begin{aligned} p - p_2 &= \zeta_2 \frac{\rho}{2}\left(\frac{A_2 u_p}{A_8}\right)^2 + \zeta_3 \frac{\rho}{2}\left(\frac{Q_i}{A_8}\right)^2 \\ &= K_2 (A_2 u_p)^2 + K_3 Q_i^2. \end{aligned} \tag{4.21}$$

(3) Pressure Differential Between the Front Chamber (p_2) and the Control Chamber (p_3)

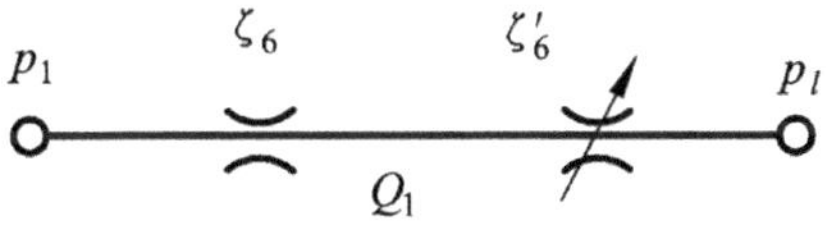

Fig. 4.6 Differential pressure calculation model between the rear chamber and valve's return chamber during the return acceleration phase

$$p_2 - p_3 = \zeta_5 \frac{\rho}{2}\left(\frac{A_5 u_v}{A_7}\right)^2 = K_5 (A_5 u_v)^2. \tag{4.22}$$

where $\left(A_7 = \frac{\pi}{4} d_7^2\right)$.

(4) Differential pressure between the rear chamber (p_1) and the valve's return chamber (p_l) is shown in Fig. 4.6.

$$\begin{aligned} p_1 - p_l &= \zeta_6 \frac{\rho}{2}\left(\frac{A_1 u_p}{A_8}\right)^2 + \Delta p_v \zeta_6 \prime \\ &= K_6 (A_1 u_p)^2 + \Delta p_v \zeta_6 \prime \end{aligned} \tag{4.23}$$

In Eq. (4.23), Δp_v represents the additional differential pressure caused by changes in the valve opening, and its calculation is detailed in Sect. 4.7. ζ_6 denotes the local resistance coefficient when the valve is fully open.

2. **Return Deceleration and Impact Acceleration Phase**

(1) Pressure Differential Between the Front Chamber (p_2) and the High-pressure Accumulator (p_h)

Similarly, Eq. (a) from the return acceleration phase can be used, but determining the flow direction is somewhat more complex.

During the return deceleration, Q_2 is still directed toward the front chamber, opposite to the direction of pressure difference ($p_2 - p_h$), so a sign "−" is applied to the first term on the right side of Eq.(a). However, during the impact stroke acceleration, Q_2 flows out of the front chamber in the same direction as the pressure difference, and in this case, a sign "+" should be applied.

The expression and direction of the flow rate Q_4 are more complicated to determine. During the return deceleration phase, the flow Q_1 in the rear chamber is also directed into the accumulator because the hydraulic impact mechanism acts as a pump at this time. The accumulator's charging flow is

$$Q_4 = Q_i + Q_1 - Q_2. \tag{b}$$

The flow direction of Q_4 aligns with the direction of the pressure differential, so the sign of the second term on the right side of Eq. (a) is positive "+".

After the return deceleration ends and transitions to the impact stroke, the flow direction of Q_1 and Q_2 into the rear and front chambers of the piston reverse, and the accumulator's charging and discharging flow rate becomes:

$$Q_4 = Q_i - (Q_1 - Q_2). \tag{c}$$

When $Q_4 > 0$, meaning that the piston requires less flow rate at the beginning of the impact stroke, Q_4 continues to be directed into the accumulator, and a '+' sign is still applied to the second term on the right side of Eq. (a).

When $Q_4 < 0$, indicating that the piston speed has increased and the required flow rate exceeds the pump supply, the accumulator will discharges to supplement at a flow rate Q_4. Since its flow direction is opposite to the direction of the pressure differential, a "−" sign is applied to the second term on the right side of Eq. (a).

By utilizing the fact that in the calculation of Q_1 and Q_2, the velocity u_p can be either positive or negative, Eqs. (b) and (c) can be unified into the following form:

$$Q_4 = Q_i + (A_1 - A_2)u_p. \qquad \text{(d)} \tag{d}$$

By introducing the sign function, equation (a) can be written as:

$$\begin{aligned} p_2 - p_h =& \text{sgn}(Q_4)\zeta_4 \frac{\rho}{2}\left(\frac{Q_4}{A_8}\right)^2 - sgn\left(u_p\right)\zeta_2 \frac{\rho}{2}\left(\frac{Q_2}{A_8}\right)^2 \\ &= sgn\left[Q_i + (A_1 - A_2)u_p\right]K_4[Q_i + (A_1 - A_2)u_p]^2 \\ &- sgn\left(u_p\right)K_2(A_2u_p)^2. \end{aligned} \tag{4.24}$$

(2) Pressure Differential Between Valve Inlet (p) and Front Chamber (p_2).

Since the flow rate Q_3 varies in both magnitude and direction during this period, further analysis is required. Referring to Fig. 4.7, as before, the pressure differential can also be written as:

$$p - p_2 = \pm\zeta_2 \frac{\rho}{2}\left(\frac{Q_2}{A_8}\right)^2 \pm \zeta_3 \frac{\rho}{2}\left(\frac{Q_3}{A_8}\right)^2. \tag{e}$$

The flow direction of Q_2 is opposite during the return deceleration and the impact stroke phases: it aligns with the pressure differential during the return stroke braking, while it is opposite to the pressure differential during the impact stroke. Therefore, a positive sign ('+') is used before the first term on the right side of equation (e) for the return deceleration, and a negative sign ('−') is used for the impact stroke.

During return deceleration, the flow direction of Q_3 is the same as the direction of the pressure differential, so the sign before the second term on the right side of equation (e) is '+'. The magnitude is:

$$Q_3 = Q_i + Q_1. \tag{f}$$

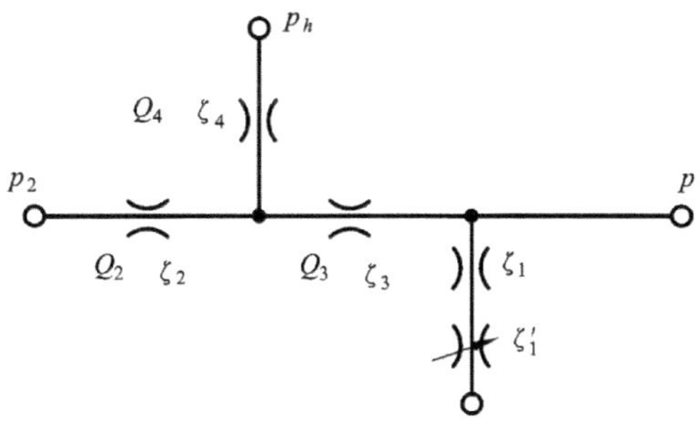

Fig. 4.7 Pressure differential calculation model between the valve inlet and front chamber during the return deceleration phase

The flow direction of Q_3 during impact stroke phase is sometimes aligned with the differential pressure and sometimes opposes to it, depending on:

$$Q_3 = Q_i - Q_1. \tag{g}$$

When $Q_3 > 0$, a '+' sign is used before the second term on the right side of the equation (e), and when $Q_3 < 0$, a '−' sign is used.

By using the sign of the piston velocity u_p , equations (f) and (g) can be unified as:

$$Q_3 = Q_i + A_1 u_p. \tag{h}$$

After introducing the sign function, equation (e) can be unified and written as:

$$\begin{aligned} p - p_2 &= sgn(u_p)\zeta_2 \frac{\rho}{2}\left(\frac{Q_2}{A_8}\right)^2 + sgn(Q_3)\zeta_3 \frac{\rho}{2}\left(\frac{Q_3}{A_8}\right)^2 \\ &= sgn(u_p)K_2(A_2u_p)^2 + sgn(Q_i + A_1u_p)K_3(Q_i + A_1u_p)^2. \end{aligned} \tag{4.25}$$

(3) Pressure Differential Between the Valve Inlet Pressure (p) and the Rear Chamber Pressure (p_1)

$$\begin{aligned} p_1 - p &= sgn(u_p)\zeta_1 \frac{\rho}{2}\left(\frac{A_1u_p}{A_8}\right)^2 + \Delta p_v(\zeta_1') \\ &= sgn(u_p)K_1(A_2u_p)^2 + \Delta p_v(\zeta_1'). \end{aligned} \tag{4.26}$$

The additional pressure differential Δp_v caused by changes in the valve opening is calculated differently during the return deceleration and impact stroke phases. Please refer to Sect. 4.7 for more details.

4.7 Influence on Rear-Chamber Pressure by Movement of Control Valve

The control valve of the hydraulic rock drill operates mostly in a fully open state, but during the reversing movement, its control edges cause variations in the valve opening area. The throttling loss caused by this change sometimes creates a significant pressure differential, even leading to intense hydraulic shocks and cavitation. When the valve spool moves to the zero position, the formation of its opening (positive opening, zero opening, negative opening) has different effects on the pressure in the rear chamber. In this book, considering the actual situation, we only address the calculation of pressure difference within the positive opening range. The fundamental calculation is referenced from the flow rate calculation equation for thin-walled orifices.

$$Q = cf\sqrt{\frac{2}{\rho}\Delta p},$$

$$\text{or} \quad \Delta p = \frac{\rho}{2} \cdot \frac{Q^2}{c^2 f^2}.$$

where:

c: flow coefficient, considering the valve ports have small rounded edges, taken as 0.7;

f: flow area of the orifice;

ρ: fluid density;

Δp: pressure differential across the orifice.

In this example, $f = \pi d_4 z$ (z represents the opening size), $Q = A_1 u_p$,

then $\Delta p = \frac{\rho}{2}\left(\frac{A_1 u_p}{c\pi d_4 z}\right)^2 = C_8\left(\frac{A_1 u_p}{z}\right)^2.$

where $C_8 = \frac{\rho}{2} \cdot \frac{1}{c^2\pi^2 d_4^2},$

or $Q = z\sqrt{\Delta p / C_8} = C_7 z\sqrt{\Delta p},$

where $C_7 = 1/\sqrt{C_8}.$

Let the stroke of the valve spool be S_v, and the positive opening size be z_0.

$$\text{Let} \quad S_{1v} = (S_v - z_0)/2,$$

$$S_{2v} = S_{1v} + z_0,$$

$$S_{3v} = S_v/2.$$

Next, we will provide two different methods for calculating the pressure difference between the rear chamber oil pressure p_1 and the valve inlet oil pressure p, and the outlet oil pressure p_l.

1. **Valve Moves in the Non-Positive Opening Range**

At this time, different calculations are required depending on whether the valve is in the return stroke or impact stroke reversal state.

(1) Return Stroke Reversal, $0 \le y_v \le S_{1v}$.

Figure 4.8 represents this state. At this time, the piston is in the final stage of the return stroke acceleration phase, and the oil in the rear chamber, $Q_1 = A_1 u_p$, enters the return oil chamber p_l through the varying opening z_1. Taking the right extreme position of the valve as the zero point for y_v, the calculation formula for the opening size is $z_1 = S_{2v} - y_v$. The formula (4.27) is the calculation formula of Δp_v in Eq. (4.23).

$$\text{At this time} \quad p_1 - p_l = \Delta p_v = C_8 \left(\frac{A_1 u_p}{S_{2v} - y_v} \right)^2, \tag{4.27}$$

which is the calculation formula of Δp_v in Eq. (4.23).

(2) Return Stroke Reversal: $S_{2v} \le y_v \le S_v$.

In Fig. 4.9, when u_v is directed towards the left, it indicates this state. Q_1 flows from p_1 to p, and at this moment, the piston has entered the deceleration phase.

The calculation formula of Δp_v in Eq. (4.26) is:

$$z_2 = y_v - s_{1v},$$

$$p_1 - p = \Delta p_v = C_8 \left(\frac{A_1 u_p}{y_v - S_{1v}} \right)^2.$$

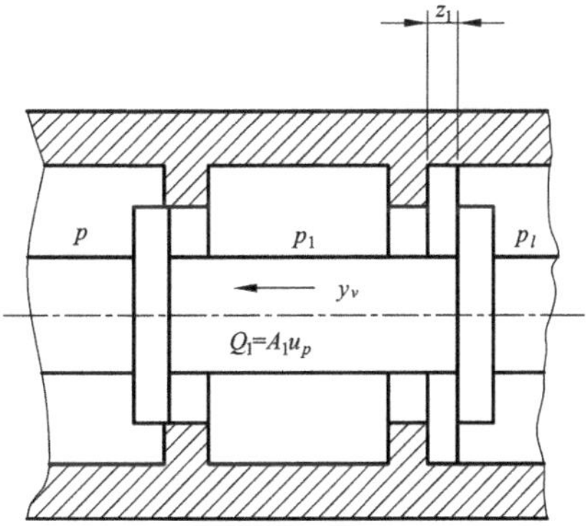

Fig. 4.8 Pressure differential calculation model for return stroke reversal ($0 \le y_v \le S_{1v}$)

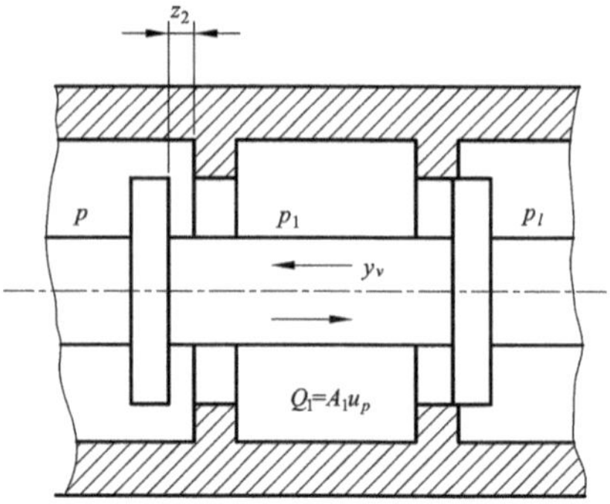

Fig. 4.9 Pressure differential calculation model during return stroke reversal ($S_{2v} \le y_v \le S_{1v}$) and impact stroke reversal

(3) Impact Stroke Reversal: $0 \le y_v \le S_{1v}$.

In Fig. 4.9, when u_v is directed to the right, it represents this state. Q_1 flows from p to p_1, and the zero point of y_v is set at the left extreme position of the valve.

The calculation formula of Δp_v in Eq. (4.26) is:

$$z_2 = S_{1v} - y_v,$$

$$p - p_1 = \Delta p_v = C_8\left(\frac{A_1 u_p}{S_{1v} - y_v}\right)^2.$$

Because the direction of pressure differential between p and p_1 in the two states (2) and (3) is opposite, Δp_v in formula (4.26) is uniformly expressed as:

$$p_1 - p = \Delta p_v = sgn(u_p) C_8\left(\frac{A_1 u_p}{z_2}\right)^2. \tag{4.28}$$

2. **The Valve Moves Within the Positive Opening Interval z0**

At this time, it results in the state where the three chambers p, p_1 and p_l are interconnected, as shown in Fig. 4.10, which is referred to as the state where the front and rear chambers simultaneously connect to the return oil. Both the valve's return stroke reversal and impact stroke reversal pass through this state. The relevant calculations are described as follows:

(1) Return Stroke Reversal

After the state shown in Fig. 4.8 ends, the system enters the state shown in Fig. 4.10. The opening amounts z_1 and z_2 in the figure can be calculated as follows:

$$z_2 = y_v - S_{1v},$$

$$z_1 = z_0 - z_2,$$

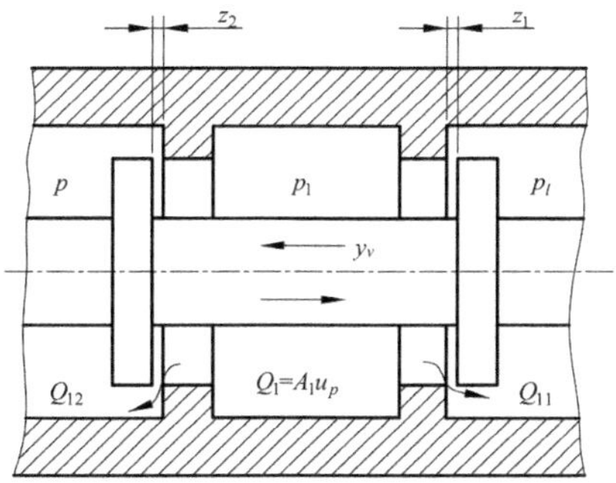

Fig. 4.10 Pressure differential calculation model of valve in positive opening interval

z_0 is the positive opening amount.

The relationship between p_1 and p, p_l is derived based on the flow balance equation:

$$Q_1 = Q_{11} + Q_{12}, \tag{a}$$

$$\text{where} \quad Q_{11} = C_7 z_1 \sqrt{p_1 - p_l}, \tag{b}$$

$$Q_{12} = sgn(p_1 - p) C_7 z_2 \sqrt{|p_1 - p|}. \tag{c}$$

The sign function $sgn(p_1 - p)$ is added before the expression of Q_{12} is because within the z_0 range, both cases of $p_1 > p$ and $p_1 < p$ are possible to occur. Therefore, we have:

$$A_1 u_p = C_7 z_1 \sqrt{p_1 - p_l} + sgn(p_1 - p) C_7 z_2 \sqrt{|p_1 - p|}. \tag{d}$$

When $p_1 > p$, the formula (d) can be rewritten as:

$$A_1 u_p = C_7 z_1 \sqrt{p_1 - p_l} + C_7 z_2 \sqrt{p_1 - p}.$$

It can be solved:

$$p_1 = \frac{2ab - p - p_l - \sqrt{(p - p_l)^2 - 4ab(p + p_l) + 4\left(b^2 + a^2 p p_l\right)}}{2\left(a^2 - 1\right)}, \tag{4.29}$$

where $a = \frac{z_1^2 + z_2^2}{2z_1 z_2}$

$$b = \frac{z_1^2 p_l + z_2^2 p + C_8 (A_1 u_p)^2}{2z_1 z_2}.$$

Specially, when $z_1 = z_2$, then:

$$p_1 = \frac{pp_l - b^2}{p + p_l - 2b}. \tag{4.30}$$

When $p_1 < p$, the formula (d) can be rewritten as:

$$A_1 u_p = C_7 z_1 \sqrt{p_1 - p_l} - C_7 z_2 \sqrt{p - p_1}.$$

It can be solved:

$$p_1 = \frac{2cd + p + p_l + \sqrt{(p - p_l)^2 + 4cd(p + p_l) - 4\left(d^2 + c^2 pp_l\right)}}{2\left(c^2 + 1\right)}, \tag{4.31}$$

where $c = \frac{z_1^2 - z_2^2}{2z_1 z_2}$,

$$d = \frac{z_1^2 p_l - z_2^2 p + C_8 (A_1 u_p)^2}{2z_1 z_2}.$$

(2) Impact stroke reversal

The impact stroke reversal occurs at the late stage of the piston impact phase, during which the piston speed is high, and consequently $Q_1 = A_1 u_p$ is large. However, the positive opening z_0 is generally small, causing the hydraulic fluid to be in a strong throttling state. Typically, when the valve has not yet entered the z_0 interval, the flow passing through z_2 cannot meet the requirements of the piston speed (Q_1). At this time, $p_1 \leq 0$. Therefore, except for the related judgment calculations, the calculation of p_1 within the z_0 interval has no practical significance.

Chapter 5
Nonlinear Model of Hydraulic Impact Mechanisms

5.1 General Survey of Basic Equations

Based on the analysis in the previous chapter, the mathematical equations related to piston, valve spool, high-pressure accumulator, and high-pressure oil flow rate balance are summarized as follows:

1. **Dynamic balance equation of piston and valve spool**

(1) Piston Equation (including all piston motion states)

$$m_p \frac{d^2 y_p}{dt^2} + B_1 \frac{dy_p}{dt} + sgn\left(\frac{dy_p}{dt}\right)\left[B_2(p_2 - p_l) + B_3(p_1 - p_l)\right] + B_4(p_2 - p_l) + B_5(p_1 - p_l) = A_2 p_2 - A_1 p_1 \tag{5.1}$$

(2) Valve Spool Equation (including all valve spool motion states)

$$m_v \frac{d^2 y_v}{dt^2} + B_6 \frac{dy_v}{dt} - B_7(p - p_l) + B_8(p - p_l) + B_9(p - p_3) = A_3(p - p_l) - A_5(p - p_3) \tag{5.2}$$

2. **High-pressure oil flow rate balance equation**

The compensatory flow rate generated by the compression or expansion of hydraulic oil and the high-pressure hose due to pressure changes is:

$$Q_e = \frac{V_k}{K_E}\frac{dp}{dt} = K_v \frac{dp}{dt} \tag{5.3}$$

(1) State A (Valve stationary; piston return stroke acceleration)

Q. He, *Research and Design of Hydraulic Impact Mechanisms*,
https://doi.org/10.1007/978-981-96-4864-1_5

$$Q_i - Q_l = A_2 \frac{dy_p}{dt} - \frac{dV_h}{dt} + \frac{V_k}{K_E} \frac{dp}{dt} \tag{5.4}$$

The last two terms on the right side of the equation represent the flow rate generated by the volume change of the accumulator and the volume change of the inlet hose and the oil within it.

(2) State B (valve return stroke reversal; piston return stroke acceleration)

$$Q_i - Q_l = A_2 \frac{dy_p}{dt} + A_3 \frac{dy_v}{dt} - \frac{dV_h}{dt} + \frac{V_k}{K_E} \frac{dp}{dt} \tag{5.5}$$

(3) State C and D (valve return stroke reversal; piston return stroke deceleration)

$$Q_i - Q_l = (A_2 - A_1) \frac{dy_p}{dt} + A_3 \frac{dy_v}{dt} - \frac{dV_h}{dt} + \frac{V_k}{K_E} \frac{dp}{dt} \tag{5.6}$$

(4) State E and F (valve stationary; piston return stroke deceleration or impact stroke acceleration)

$$Q_i - Q_l = (A_2 - A_1) \frac{dy_p}{dt} - \frac{dV_h}{dt} + \frac{V_k}{K_E} \frac{dp}{dt} \tag{5.7}$$

(5) State G (valve impact stroke reversal; piston impact stroke acceleration)

$$Q_i - Q_l = (A_1 - A_2) \frac{dy_p}{dt} + (A_5 - A_3) \frac{dy_v}{dt} - \frac{dV_h}{dt} + \frac{V_k}{K_E} \frac{dp}{dt} \tag{5.8}$$

(6) State H (valve switching the oil circuit in advance; piston impact stroke deceleration)

$$Q_i - Q_l = A_2 \frac{dy_p}{dt} + (A_5 - A_3) \frac{dy_v}{dt} - \frac{dV_h}{dt} + \frac{V_k}{K_E} \frac{dp}{dt} \tag{5.9}$$

(7) State I (valve impact stroke reversal; piston pause)

$$Q_i - Q_l = (A_5 - A_3) \frac{dy_v}{dt} - \frac{dV_h}{dt} + \frac{V_k}{K_E} \frac{dp}{dt} \tag{5.10}$$

(8) State I′ (both valve and piston pause)

$$Q_i - Q_l = -\frac{dV_h}{dt} + \frac{V_k}{K_E} \frac{dp}{dt} \tag{5.11}$$

(9) State A′ (valve continues impact stroke reversal; piston return stroke acceleration)

$$Q_i - Q_l = A_2\frac{dy_p}{dt} + (A_5 - A_3)\frac{dy_v}{dt} - \frac{dV_h}{dt} + \frac{V_k}{K_E}\frac{dp}{dt} \tag{5.12}$$

3. **High-pressure Accumulator Pressure**

$$p_h = p_{ah}\left(\frac{V_{ah}}{V_h}\right)^k = C_h/V_h^k \tag{5.13}$$

5.2 Calculation of Leakage and Pressure Differential

1. **Leakage calculation**

$$\begin{aligned} Q_l &= Q_{l1} + Q_{l2} + Q_{lv} \\ &= R_1(p_1 - p_l) + R_3(p_2 - p_l) + sgn(u_p)(R_2 + R_4)u_p \\ &+ R_5(p - p_l) \end{aligned} \tag{5.14}$$

2. **Pressure Differential Calculation**

(1) Return Stroke Acceleration

$$p_2 - p_h = sgn(Q_i - A_2u_p)K_4(Q_i - A_2u_p)^2 - K_2Q_2^2 \tag{5.15}$$

$$p - p_2 = K_2Q_2^2 + K_3Q_i^2 \tag{5.16}$$

$$p_2 - p_3 = K_5Q_5^2 \tag{5.17}$$

(It is only used when the valve is reversed during the return stroke, that is, it is used for the *B*, C and D states).

$$p_1 - p_l = K_6Q_1^2 + C_8\left(\frac{Q_1}{S_{2v} - y_v}\right)^2 \tag{5.18}$$

(2) Return Stroke Deceleration and Impact Stroke Acceleration

$$\begin{aligned} p_2 - p_h = {} & sgn[Q_i + (A_1 - A_2)u_p]K_4[Q_i + (A_1 - A_2)u_p]^2 \\ & -sgn(u_p)K_2(A_2u_p)^2 \end{aligned} \tag{5.19}$$

$$p - p_2 = sgn(u_p)K_2(A_2u_p)^2 + sgn(Q_i + A_1u_p)K_3(Q_i + A_1u_p)^2 \quad (5.20)$$

$$p_1 - p = sgn(u_p)K_1Q_1^2 + sgn(u_p)C_8\left(\frac{A_1u_p}{S_{1v} - y_v}\right)^2 \quad (5.21)$$

(3) Positive Opening Interval z_0

State C: When the valve undergoes return stroke reversal and enters the positive opening interval z_0, the pressure p_1 in the rear chamber is as described in Sect. 4. 7:

- when $p_1 > p$, replace Eq. (5.21) with Eq. (4.29) or Eq. (4.30) (when $z_1 = z_2$);
- when $p_1 < p$, replace Eq. (5.21) with Eq. (4.31).

Other pressure differential calculations are the same as during the return deceleration or impact stroke acceleration phases.

State H: When the valve undergoes impact stroke reversal into the positive opening interval z_0, the piston decelerates during impact stroke. In fact, at this time, $p_1 < 0$, and we can let:

$$\boldsymbol{p}_1 = 0 \quad (5.22)$$

5.3 Analysis of Return Oil Motion of Hydraulic Impact Mechanisms

Whether it is a rear-control type or a dual- control type hydraulic impact mechanism, there is always a stage in each work cycle when oil is not discharged into the return oil pipe, except for leakage. This is especially true in rear-control type impact mechanisms, where a large amount of oil is discharged into the return oil pipe only during the piston's return stroke acceleration phase. Moreover, the oil discharge varies from small to large as the piston speed increases, resulting in significant pulsation in the return oil pipe. In this section, only the return oil motion state with a return accumulator installed is analyzed.

Figure 5.1 visually illustrates the changes in the velocity of the oil column in the return oil pipe, the charging and discharging process of the return oil accumulator, and the oil return pressure (i.e. the accumulator pressure) within one working cycle for the rear-control type hydraulic impact mechanism.

The curves in Fig. 5.1 indicate that the return oil accumulator is only charged during the piston's return stroke acceleration phase. During the return stroke deceleration and impact stroke acceleration, the accumulator discharges to maintain the continuity of flow in the pipe. Additionally, when the oil pressure p_f generated by

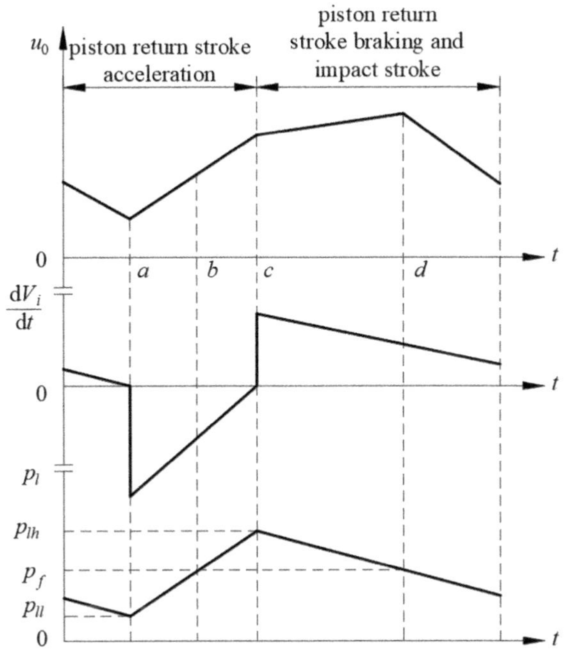

Fig. 5.1 Oil column velocity curve, return accumulator charging/discharging curve, and return oil pressure variation curve

the along-path resistance of the return oil column and throttle resistance such as the oil filter is lower than the accumulator pressure p_l, the oil column accelerates. Conversely, when p_f exceeds p_l, the oil column decelerates.

Based on the above analysis, we proceed to establish the mathematical model of the return oil system.

Here, the mass of the oil column in the return oil pipe, denoted as m_0, is treated as a lumped parameter. The resistance it experiences during motion comes from both the along-path resistance and throttling resistance caused by the oil filter at the end of the pipe. These two resistance components are combined and represented as the back pressure of the return oil, denoted as p_f.

Since the flow in the return pipe is essentially a highly pulsating, unsteady flow, the along-path pressure loss p_{f1} is calculated by treating it as turbulent flow in a rough pipe.

$$p_{f1} = \lambda \rho \frac{l_0}{d_0} \cdot \frac{u_0^2}{2} = K_\lambda u_0^2$$

$$\lambda = \left[2lg\left(\frac{d_0}{\Delta}\right) + 1.14 \right]^{-2}$$

In the formula, Δ represents the absolute roughness of the hose, with $\Delta = 30 \times 10^{-6}$m.

How to calculate the back pressure p_{f2} formed by the throttling of the oil filter? The rated pressure drops p_d of an oil filter with a rated flow rate of Q_d is known in advance. Using the formula.

$Q_d = C_k\sqrt{p_d}$,

the throttling coefficient can be obtained as:

$$C_k = \frac{Q_d}{\sqrt{p_d}}$$

Thus, the throttling back pressure at a certain return speed can be calculated as:

$$p_{f2} = \frac{Q_0^2}{C_k^2} = \frac{(A_0 u_0)^2}{C_k^2} = K_c u_0^2$$

The total return back pressure formed by the two is

$$p_f = p_{f1} + p_{f2} = (K_\lambda + K_c)u_0^2,$$

$$F_{sc} = A_0 p_f = A_0(K_\lambda + K_c)u_0^2 = K_{10}u_0^2 \tag{5.23}$$

Due to the relatively small oil discharge from the valve spool, it is not considered in the equations for the return oil system. The equations for the two main stages of the return oil system are as follows:

1. **Return Stroke Acceleration Phase**

Dynamic balance equation for oil column m_0:

$$m_0\frac{d^2y_0}{dt^2} + K_{10}\left(\frac{dy_0}{dt}\right)^2 = A_0 p_l \tag{5.24}$$

Return oil flow balance equation:

$$A_1\frac{dy_p}{dt} + Q_l = A_0\frac{dy_0}{dt} - \frac{dV_l}{dt} \tag{5.25}$$

Return oil accumulator pressure:

$$p_l = p_{al}\left(\frac{V_{al}}{V_l}\right)^k = C_l/V_l^k \tag{5.26}$$

2. **Return Stroke Braking and Impact Stroke Acceleration Phase**

The dynamic balance equation for the return oil column m_0 and the equation for the return oil accumulator are the same as those during the return stroke acceleration. Since the piston is no longer discharging oil, the flow balance equation is.

$$Q_l = A_0 \frac{dy_0}{dt} - \frac{dV_l}{dt} \tag{5.27}$$

5.4 Conclusion of Mathematic Model

In the previous sections, the mathematical equations describing the motion characteristics of the piston, valve, high-pressure accumulator, and return oil system were summarized. Based on this, and in conjunction with the previous analysis, the mathematical model reflecting the movement of 12 operating states (A, A′, B, C, etc.) within one working cycle of the hydraulic impact mechanism is summarized in Table 5.1.

The state A″ is a very abnormal operating condition; it should be avoided as much as possible. Therefore, in the simulation calculations, if it is determined that this state occurs, it will be simplified before further simulation and solving.

Table 5.1 Mathematical model of 12 working states of hydraulic impact mechanisms

Status codes and characteristics	The formula number of the mathematical model	Model code
A Valve spool stationary, piston return stroke acceleration	(5.1), (5.4), (5.13), (5.14), (5.15), (5.16), (5.18), (5.23), (5.24), (5.25), (5.26)	mA
A′ Valve impact stroke reversal, piston return stroke acceleration	(5.1), (5.2), (5.12), (5.13), (5.14), (5.15), (5.16), (5.18), (5.23), (5.24), (5.25), (5.26)	mA′
B Valve return stroke reversal, piston return stroke acceleration	(5.1), (5.2), (5.5), (5.13), (5.14), (5.15), (5.16), (5.17), (5.18), (5.23), (5.24), (5.25), (5.26)	mB
C Valve in positive opening interval, piston return stroke braking	(5.1), (5.2), (5.6), (5.13), (5.14), (5.17), (5.19), (5.20), (5.21), (4.29), (4.30), (4.31), (5.23), (5.24), (5.26), (5.27)	mC
D Valve return stroke reversal, piston return stroke braking	(5.1), (5.2), (5.6), (5.13), (5.14), (5.17), (5.19), (5.20), (5.21), (5.23), (5.24), (5.26), (5.27)	mD
E Valve stationary, piston return stroke braking	(5.1), (5.7), (5.13), (5.14), (5.19), (5.20), (5.21), (5.23), (5.24), (5.26), (5.27)	mE
F valve stationary, piston impact stroke acceleration	Same as E	
G Valve impact stroke reverse, piston impact stroke acceleration	(5.1), (5.2), (5.8), (5.13), (5.14), (5.19), (5.20), (5.21), (5.23), (5.24), (5.26), (5.27)	mG
H Valve switching the oil circuit in advance, piston impact stroke braking	(5.1), (5.2), (5.9), (5.13), (5.14), (5.19), (5.20), (5.21), (5.22), (5.23), (5.24), (5.26), (5.27)	mH
H′ Valve stationary, piston braking	(5.1), (5.9), (5.13), (5.19), (5.20), (5.21), (5.23), (5.24), (5.26), (5.27)	mH′
I Valve impact stroke reversal, piston paused	(5.2), (5.10), (5.13), (5.14), (5.19), (5.20), (5.21), (5.23), (5.24), (5.26), (5.27)	mI
I′ Both valve and piston paused	(5.11), (5.13), (5.14), (5.15), (5.20)	mI′

Chapter 6
Computer Simulation of Impact Mechanisms

6.1 Characteristics of Simulation Computation of Impact Mechanisms

1. **Frequent State Transitions**

As previously mentioned, the operating states of the impactor can be broadly divided into at least four categories, with finer divisions reaching more than ten. Clearly, these frequent state transitions present certain challenges for simulation and solution, as each transition involves issues such as state recognition, state initialization, and errors arising from the transition itself. For example, the structure of a specific impact mechanism dictates that the control oil circuit for valve reversal is engaged when the piston has traveled 2 cm during its return stroke. However, most simulation programs can only detect when the piston's return stroke exceeds 2 cm, at which point the state transition begins. Clearly, the range of this 'exceeding' determines the error in state transitions, and as the number of state transitions increases, the accumulated error becomes larger.

2. **Multiple Cycles to Reach a Stable Solution**

Unlike general simulation solving, the simulation process for the impact mechanism is cyclical. It reflects the actual machine's transition from startup to normal operation. The hydraulic pump outputs a constant flow rate Q_i, which is injected into the impact mechanism through high-pressure hoses. The hydraulic fluid experiences inertial resistance and other resistances from the moving components (mainly the piston and valve), leading to an increase in oil pressure. The elevated oil pressure propels the moving parts to accelerate. As each working cycle continues, the average speed of the piston and other moving components increases, resulting in more flow rate consumption. Excess hydraulic fluid before stabilization accumulates in the high-pressure accumulator. Therefore, the author believes that the physical essence of the stable solution in the numerical simulation of the impact mechanism is that, in a certain working cycle, the supply flow rate from the hydraulic pump is precisely

Q. He, *Research and Design of Hydraulic Impact Mechanisms*,
https://doi.org/10.1007/978-981-96-4864-1_6

equal to the average flow rate required for the movement of the piston and valve, along with the average leakage flow rate. This can be expressed mathematically as:

$$Q_i = \overline{Q}_p + \overline{Q}_v + \overline{Q}_l.$$

It is evident that, in the cyclic process that satisfies this relationship, no matter how the volume of the gas chamber of the high-pressure accumulator changes during the cycle, it should be equal at the beginning and end of each cycle, i.e.:

$$V_{start} = V_{end}(V_s = V_e).$$

It is appropriate to use this equation as the criterion for determining whether the cyclic solving process has reached a steady state, as it reflects the physical essence of the actual machine. Before this criterion is satisfied, the gas chamber volume of the high-pressure accumulator at the end of each cycle will be smaller than at the end of the previous cycle (due to the excess hydraulic fluid accumulating from the pump's supply flow rate). As a result, the system pressure, impact energy, and impact frequency increase, but the magnitude of the increase becomes smaller and eventually approaches a stable value. The correct numerical simulation process should follow this pattern of variation.

6.2 Determination of the Hydraulic Impact Mechanism's Working State

Since there are multiple working states within a single working cycle of the impact mechanism, determining the criteria for each state is a primary issue to be addressed in simulation calculations and for achieving state transitions. Some of the identification (or judgment) criteria, for state transitions are easy to determine, while others are more complex. In the following analysis, 'A → B,' 'B → C,' etc., represent the transition from state A to state B, from state B to state C, and so on.

First, the criteria for each state transition are listed:

$$A \to B : y_p \geq S_c; \quad B \to C : y_v \geq S_{1v}; \quad C \to D : y_v \geq S_{2v};$$
$$D \to E : y_v \geq S_v; \quad E \to F : u_p \leq 0; \quad F \to G : y_p \leq S_{ic} - S_p;$$
$$G \to H : \omega_p \geq 0; \quad H \to H' : \omega_p \geq 0, y_v > S_v, y_p < -S_p;$$
$$H \to I : y_p \leq -S_p; \quad G \to I' : y_p \leq -S_p, y_v \geq S_v;$$
$$I \to A : t_s \geq t_{st}, y_v \geq S_v; \quad I \to A' : t_s \geq t_{st}, S_{2v} < y_v \leq S_v;$$
$$I \to A'' : t_s \geq t_{st}\ ,\ y_v \leq S_{2v}\ ; \quad I' \to A : t_s \geq t_{st}$$

Criterion for ending the simulation:

$$|V_s - V_e| \leq \Delta V$$

Where:

- V_s and V_e are the volumes of the high-pressure accumulator at the beginning and end of a certain working cycle, respectively.
- ΔV is the allowable difference between V_s and V_e, typically taken as 0.1% to 0.3% of the accumulator's inflation volume V_{ah}.

Notes:

(1) ω_p: piston acceleration; positive during the return acceleration phase, and negative during the return deceleration and impact acceleration phases;
(2) t_s : the time the piston pauses after impact ($t_s < t_{st}$);
(3) $G \rightarrow H$: identification of the valve's early reversing during the impact stroke phase.

The identification of G → H is somewhat more complex. The following analysis explains why $\omega_p \geq 0$ is used as the discriminant for the transition from state G to state H.

Referring to Sect. 4.7 and Fig. 4.9, during the impact phase, the valve port z_2, which controls the rear chamber, gradually narrows during the reversal process of the impactor's control valve. This increases the throttling loss at the valve port, resulting in a decrease in the rear chamber oil pressure. The maximum possible flow rate into the rear chamber (i.e., the flow rate that can pass through the valve port when the oil pressure in the rear chamber is zero) also decreases. If the feedback hole position of the impactor control valve is improperly designed, the following two phenomena may occur:

① The flow rate through the valve port z_2 cannot meet the requirements of the rear chamber, resulting in the rear chamber oil pressure $p_1 \leq 0$;
② The oil pressure in the rear chamber drops to a level that it cannot sustain the piston's acceleration, i.e., $p_1 A_1 - p_2 A_2 - f < 0$ (where f is the other resistance).

Both of these phenomena will cause the valve to switch oil circuit prematurely, reducing the impact speed. But which condition occurs first? This is a state identification problem that must be solved during the digital simulation of the impact mechanism.

At this point, the oil pressure in the front chamber p_2 is higher than the valve inlet oil pressure p.

Let

$$p_2 = p + \Delta p.$$

At this point, the flow rate required by the rear chamber is:

$$Q_1 = A_1 u_p.$$

The maximum possible flow rate through valve port z_2 is:

$$Q_{\max} = C_7 z_2 \sqrt{p} = C_7 z_2 \sqrt{p_2 - \Delta p}.$$

The condition that the first phenomenon will not occur is:

$$Q_{\max} > Q_1,$$

that is,

$$C_7 z_2 \sqrt{p_2 - \Delta p} > A_1 u_p,$$

$$p_2 > \left(\frac{A_1 u_p}{C_7 z_2}\right)^2 + \Delta p = m. \tag{a}$$

On the other hand, the pressure differential between the rear chamber pressures p_1 and p caused by throttling of the valve port is:

$$p - p_1 = \left(\frac{A_1 u_p}{C_7 z_2}\right)^2,$$

that is,

$$p_1 = p_2 - \Delta p - \left(\frac{A_1 u_p}{C_7 z_2}\right)^2.$$

The condition for the second phenomenon not to occur is:

$$A_1 p_1 - A_2 p_2 > 0 \text{ (neglecting other resistances } f\text{)},$$

that is,

$$A_1 \left[p_2 - \Delta p - \left(\frac{A_1 u_p}{C_7 z_2}\right)^2 \right] - A_2 p_2 > 0,$$

or

$$p_2 > \frac{A_1}{A_1 - A_2} \left[\left(\frac{A_1 u_p}{C_7 z_2}\right)^2 + \Delta p \right] = \frac{A_1}{A_1 - A_2} \cdot m > m. \tag{b}$$

By comparing Eqs. (a) and (b), it is evident that the front chamber pressure required by Eq. (b) is higher than that required by Eq. (a). Clearly, the second phenomenon will occur first. In other words, $p_1 A_1 - p_2 A_2 - f < 0$ can be used as the discriminant for state identification. Alternatively, the piston acceleration in the later stage of the impact stroke, $\omega_p \geq 0$, can be used as the criterion for the valve to switch the oil circuit early during the impact stroke phase ($G \rightarrow H$). $\omega_p \geq 0$ is chosen because, during the impact stroke phase, the piston acceleration has already taken on a negative value compared to the return acceleration phase. As soon as ω_p changes sign, it indicates that the impact stroke reversal has occurred early.

(4) Determination of the Actual End of the Return Stroke Acceleration Phase

In the previous chapters, the valve's reversal during the return stroke into the positive opening interval was considered the indication of the end of the return acceleration phase, marking the transition from state *B* to state *C*. However, in reality, before this moment, the pressure p_1 increases due to the oil flow from the rear chamber being affected by throttling, causing the force accelerating the piston to drop to zero. Therefore, the criterion for determining the true end of the return acceleration phase is $\omega_p \leq 0$.

6.3 Pua Quasi-Constant Acceleration Computation and Correction Computation

The characteristics of computational numerical simulation of hydraulic impact mechanisms were discussed in the previous two sections. Given the need for multiple cycles to achieve stable solutions and the frequent state transitions within a single operating cycle during the simulation process, it is necessary to choose a suitable numerical calculation method and minimize the accumulation of errors during state transitions. To address this, the author proposes a simplified quasi-constant acceleration numerical method and introduces correction calculations during state transitions.Consider a second-order differential equation

1. **Quasi-Constant Acceleration Numerical Method (PUA Method)**

In Sects. 6.1 and 6.2, mathematical models for the various operational states of the hydraulic impact mechanism were established. These models consist of nonlinear systems of differential equations with complex initial conditions. Unless the model is significantly simplified and linearized, which would result in considerable deviation from actual operating conditions, an analytical solution is often unattainable. Therefore, numerical methods for solving differential equations must be employed.

The current methods for obtaining numerical solutions to differential equations are diverse, including the Euler method, the improved Euler method, and the Runge–Kutta method. Among these, the fourth-order Runge–Kutta method, known for its higher accuracy, is widely utilized. In this book, based on the characteristics of the mathematical model of the hydraulic impact mechanism, a numerical computation method for solving second-order differential equations with quasi-constant acceleration is proposed. The basic idea is to treat the acceleration of an object undergoing variable acceleration as constant over a suitably short time interval, and then use the relevant formulas for uniformly accelerated motion to calculate the velocity and displacement of the object after experiencing this time interval. In this context, the second derivative of the second-order differential equation is assigned the physical meaning of "acceleration," and the suitably short time interval is referred to as the computational step size.

Consider a second-order differential equation:

$$\frac{d^2x}{dt^2} = f\left(\frac{d_x}{dt}, x, t\right).$$

Let the acceleration be:

$$\omega = \frac{d^2x}{dt^2},$$

$$F = f\left(\frac{d_x}{dt}, x, t\right).$$

Assuming the computational step size is Δt, and the initial values of velocity and displacement are u_0 and x_0, respectively, then after the time step Δt, we have:

$$\omega = F, \tag{6.1}$$

$$u = u_0 + \omega \Delta t, \tag{6.2}$$

$$x = x_0 + u_0 \Delta t + \frac{1}{2}\omega \Delta t^2. \tag{6.3}$$

Equations (6.1)–(6.3) are the iterative formulas for the quasi-constant acceleration numerical method for solving second-order differential equations. Its accuracy is naturally lower than that of the fourth-order Runge-Kulla method, but the computational effort is much less, especially when it is given a clear physical meaning, which greatly facilitates programming. Additionally, it makes the correction calculations introduced during state transitions easier to implement.

2. **Correction Calculation During State Transition**

The criteria for the 12 state transitions within one working cycle of the impact mechanism were listed in Sect. 6.2, all expressed in the form of inequalities. For example, $y_p \geq S_c$ for the transition from A to B, and $u_p \leq 0$ for the transition from E to F. During the iterative calculation, except for occasional cases, the state transitions generally occur when $y_p > S_c$ or $u_p < 0$, introducing additional judgment errors. Furthermore, with as many as a dozen state transitions occurring frequently in one cycle, the resulting cumulative errors cannot be ignored. Therefore, based on the quasi-constant acceleration numerical method proposed above, the author has devised the following correction calculation method.

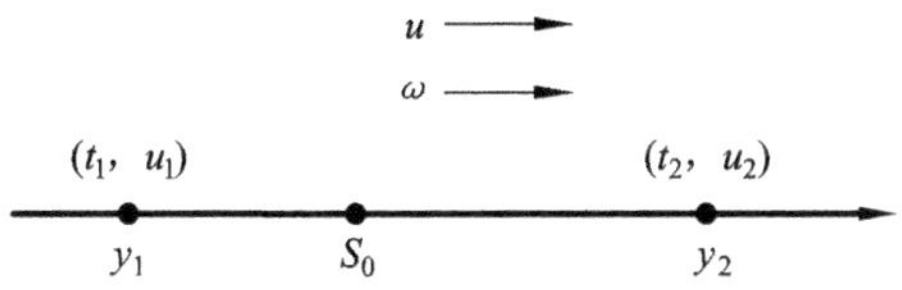

Fig. 6.1 Correction calculation diagram for y > S0

(1) Correction Calculation After $y > S_0$ is Satisfied:

As shown in Fig. 6.1, a piston or valve spool with an acceleration of ω moves from point y_1 to y_2 within a given time step Δt, reaching point y2 at time t_2 with velocity u_2. At this point, it is found that the discriminant $y_2 > S_0$ holds true. If a state transition is made at this moment, several errors will occur in the standard parameter values relative to the judgment criterion position S_0.

Displacement error: $y_2 - S_0$.

Velocity Error: $u_2 - u_0$.

Time Error: $t_2 - t_0$.

If the values of u_0, t_0, and other parameters at the S_0 point can be determined, these errors can be eliminated. Based on the basic formulas of uniformly accelerated motion:

$$u_2^2 - u_0^2 = 2\omega(y_2 - S_0)$$

$$t_2 - t_0 = (u_2 - u_0)/\omega,$$

we can obtain:

$$u_0 = \sqrt{u_2^2 - 2\omega(y_2 - S_0)}, \tag{6.4}$$

$$t_0 = t_2 - (u_2 - u_0)/\omega, \tag{6.5}$$

$$y_0 = S_0. \tag{6.6}$$

Formulas (6.4)–(6.6) represent the velocity and corresponding time that the piston or spool should have when located at the judgment criterion position. These are the correction calculation formulas used after the discriminant expression $y > S_0$ is satisfied.

(2) Correction Calculation After $u < 0$ is Satisfied

The piston, decelerating with negative acceleration ω, should reduce its velocity from $u_1 > 0$ to zero at point y_0 in Fig. 6.2. However, in simulation calculations, it is often the case that after a time step Δt, the piston begins decelerating from point y_1, passes

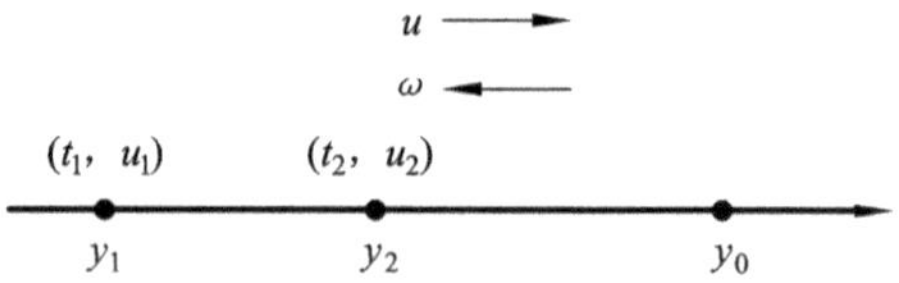

Fig. 6.2 Correction calculation diagram for u < 0

point y_0, and then accelerates in the opposite direction to point y_2. At this point, $u_2 < 0$ holds true. If a state transition ($E \rightarrow F$) is made at this moment, errors will occur.

According to the calculation formulas for constant acceleration motion:

$$t_2 - t_0 = u_2/\omega,$$

$$y_2 - y_0 = \frac{1}{2}\omega(t_2 - t_0)^2,$$

We can derive the correction formula after $\mathrm{u_p} < 0$ holds true:

$$t_0 = t_2 - u_2/\omega, \tag{6.7}$$

$$y_0 = y_2 - \frac{1}{2}\omega(t_2 - t_0)^2, \tag{6.8}$$

$$u_0 = 0. \tag{6.9}$$

Using formulas (6.7) and (6.8), the time and displacement when the piston velocity $u_p = 0$ can be obtained.

(3) Correction Calculation When the Piston and Spool Move Simultaneously

The Corrective calculation methods for the separate movements of the piston and spool have been provided earlier. When they move simultaneously, if one of them, such as the piston, first crosses a certain state transition criterion position S_0, it is necessary to correct not only the relevant parameters of the piston but also the corresponding parameters of the spool.

Suppose that at the moment the piston crosses S_0, its time, acceleration, velocity, and displacement are t_{p2}, ω_p, u_{p2}, and y_{p2}, respectively. The corresponding parameters of the piston at the standard position can be represented by the following correction formulas:

$$\Delta y_p = y_{p2} - S_0, \tag{6.10}$$

$$u_{p0} = \sqrt{u_{p2}^2 - 2\omega_p \Delta y_p}, \tag{6.11}$$

$$\Delta t_p = \frac{(u_{p2} - u_{p0})}{\omega_p}. \tag{6.12}$$

At this time, the correction formulas for the spool's velocity and displacement are:

$$u_{v0} = u_{v2} - \omega_v \Delta t_p \tag{6.13}$$

$$\Delta y_v = \frac{u_{v2}^2 - u_{v0}^2}{2\omega_v}, \tag{6.14}$$

$$y_{v0} = y_{v2} - \Delta y_v. \tag{6.15}$$

In these formulas, the subscript with 'v' represents the corresponding parameter values of the valve spool.

Through the above analysis, the basic formulas for three types of correction calculations during state transitions in the simulation process have been derived. In practice, it is also necessary to perform correction calculations for the accumulator volume and working oil pressure. The subsequent analysis will demonstrate that these correction calculations are essential for the computer simulation of hydraulic impact mechanisms. Without them, the reliability of the simulation results will be significantly reduced.

3. **Effectiveness of the PUA Quasi-Uniform Acceleration Numerical Method with Correction Calculations**

In order to intuitively understand the effect of this method, we simulate the mathematical model of piston movement of the rear-controlled hydraulic impact mechanism under the ideal state described by equations:

$$\begin{cases} m_p \frac{d^2 y_p}{dt^2} = Ap \\ Q_i = A\frac{dy_p}{dt} - \frac{dV_h}{dt} \\ p = C_h / V_h^k, \end{cases}$$

which does not consider valve movement, leakage and other losses, and does not consider the compressibility of oil.

The equations can be combined into one nonlinear differential equation:

$$m_p \frac{d^2 y_p}{dt^2} = AC_h / (V_0 + Ay_p - Q_i t)^k.$$

where, V_0 represents the initial volume of the high-pressure accumulator. The value of A is determined as follows: when the piston is in the return acceleration phase, $A = A_2$; when the piston is in the return deceleration and impact stroke phases, $A = A_2 - A_1$. The meanings of the other symbols are as previously defined.

The input calculation values are:

$$A_2 = 2 \times 10^{-4}\mathrm{m}^2, A_1 = 6 \times 10^{-4}\mathrm{m}^2,$$

$$m_p = 4.5\,\mathrm{kg}\;, Q_i = 0.6 \times 10^{-3}\mathrm{m}^3/\mathrm{s}\;.$$

In the formula $C_h = p_{ah} V_{ah}^k$, let $P_{ah} = 7 \times 10^6$ Pa, $V_{ah} = 0.1 \times 10^{-3}$ m^3.

The stroke of the piston at the end of the return acceleration phase is specified as $S_r = 0.02$ m.

Digital simulation programs for the above-mentioned impact mechanism were developed using both the fourth-order Runge–Kutta method (abbreviated as R-K method) and the PUA method (combined with correction calculations). The simulation results are recorded in Tables 6.1 and 6.2. The stability criterion for the simulation is based on the difference in the accumulator gas chamber volume from the beginning to the end of one cycle, as proposed in Sect. 6.1, specifically:

$$|V_s - V_e| \leq \Delta V.$$

In this case, $\Delta V = 1.0^{-8}(\mathrm{m}^3) = 0.01(\mathrm{mL})$.

The meanings of the symbols in the table header are as follows:

H_0: calculation step size, ranging from 0.02 ms to 0.6 ms.

F: impact frequency, calculated based on the accumulation of all steps in one cycle during the simulation:

$$\mathrm{F} = 1/\sum^{\mathrm{T}} H_0$$

F′: impact frequency, calculated using the formula.

$$F' = \frac{Q_i}{S_p(A_1 - A_2)}$$

For a hydraulic impact mechanism with no flow losses, the piston only consumes high-pressure fluid during the impact stroke, while during the return stroke, the accumulator stores energy without consuming energy. Therefore, the piston stroke S_p and the input flow rate Q_i obtained from the simulation can be used to calculate the frequency (where T is the impact cycle).

E: kinetic energy of the piston obtained from the simulation, calculated by $E = \frac{1}{2} m_p u_p^2$.

E': The impact energy calculated based on the hydraulic energy input over one cycle of the hydraulic impact mechanism, obtained with the numerical integration formula:

Table 6.1 Digital simulation results with the R-K method

H_0/ms	F/Hz	F′/HZ	E/J	E′/J	S/mm	V/mL
0.02	49.55	50.8	145.18	146.01	29.52	71.58
0.04	49.6	50.79	144.3	145.86	29.53	71.57
0.06	49.31	50.58	143.06	145.32	29.66	72.05
0.08	49.02	51.25	140.72	143.51	29.27	72.98
0.1	48.08	50.66	140.04	143.61	29.61	73.98
0.12	48.73	51.46	138.29	142.42	29.15	73.65
0.14	47.94	51.18	137.62	152.5	29.31	74.51
0.16	47.35	50.07	136.31	141.67	29.96	75.46
0.18	47.89	50.16	136.69	142.89	29.9	74.38
0.2	47.62	50.2	136.16	143.07	29.88	74.63
0.22	46.38	49.96	132.37	139.32	30.03	77.45
0.24	46.3	51.03	132.13	139.9	29.4	77.35
0.26	46.9	50.04	132.09	140.54	29.98	76.34
0.28	45.79	50.92	128.21	136.6	29.46	79.22
0.3	45.66	49.74	128.21	137.92	30.16	78.84
0.32	46.64	52.56	127.67	136.76	28.54	78.15
0.34	43.9	49.87	126.09	135.82	30.08	82
0.36	44.8	49.7	126.67	137.37	30.18	80.14
0.38	46.17	49.8	127.27	139.18	30.12	77.67
0.4	45.45	48.97	127.54	141.41	30.63	77.65
0.42	44.92	51.51	121.72	133.51	29.12	81.52
0.44	42.88	49.25	121.65	133.42	30.46	84.38
0.46	42.63	48.67	123.37	136.4	30.82	83.48
0.48	43.4	48.43	121.79	135.04	30.97	82.91
0.5	46.51	49.99	122.04	137.22	30	77.92
0.52	45.79	53.82	120.76	133.77	27.87	80.44
0.54	44.09	52.73	116.93	131.56	28.45	83.44
0.56	42.52	50.91	116.86	131.48	29.46	85.72
0.58	41.05	49.22	116.8	131.4	30.48	87.97
0.6	42.74	49.93	114.57	129.51	30.04	86.23

$$E' = \int^{T} pQ_i dt \approx \sum^{T} pQ_i H_0.$$

Clearly, the output (impact) energy of an ideal impact mechanism should be equal to the input energy.

Table 6.2 Digital simulation results with the PUA method

H_0/ms	F/Hz	F'/Hz	E/J	E'/J	S/mm	V/ml
0.02	49.97	50.85	146.77	146.79	29.5	70.9
0.04	49.97	50.85	146.85	146.81	29.5	70.9
0.06	49.96	50.84	146.91	146.82	29.5	70.9
0.08	49.96	50.84	146.98	146.83	29.51	70.9
0.1	49.96	50.83	147.08	146.85	29.51	70.89
0.12	49.96	50.82	147.25	146.9	29.52	70.88
0.16	49.96	50.81	147.33	146.92	29.52	70.87
0.18	49.96	50.81	147.42	146.94	29.52	70.86
0.2	49.96	50.8	147.51	146.96	29.53	70.86
0.22	49.96	50.79	147.59	146.98	29.53	70.85
0.24	49.96	50.79	47.68	147.01	29.53	70.85
0.26	49.96	50.78	147.77	147.02	29.54	70.84
0.28	49.96	50.78	147.85	147.04	29.54	70.83
0.3	49.96	50.77	147.94	147.07	29.54	70.82
0.32	49.96	50.77	148.03	148.08	29.55	70.82
0.34	49.96	50.76	148.13	147.14	29.55	70.81
0.36	49.96	50.75	148.2	147.14	29.55	70.8
0.38	49.96	50.75	148.28	147.16	29.56	70.79
0.4	49.96	50.74	148.38	147.19	29.56	70.79
0.42	49.96	50.74	148.48	147.22	29.56	70.78
0.44	49.96	50.73	148.56	147.18	29.57	70.77
0.46	49.96	50.73	148.62	147.22	29.57	70.77
0.48	49.96	50.72	148.72	148.27	29.57	70.76
0.5	49.96	50.72	148.81	147.25	29.58	70.75
0.52	49.96	50.71	148.89	147.31	29.58	70.75
0.54	49.96	50.7	148.89	147.26	29.58	70.74
0.56	49.96	50.7	149.05	147.36	29.59	70.73
0.58	49.96	50.69	149.15	147.33	29.59	70.72
0.6	49.96	50.69	149.25	147.41	29.59	70.72

V: the volume of the gas chamber of the accumulator at the end of each impact.

By analyzing and comparing the data in Tables 6.1 and 6.2, it can be concluded that:

(1) The PUA Method Has Higher Accuracy Than the R-K Method

The results obtained using the PUA method are more accurate than those from the R-K method, and the former is much less affected by the calculation step size compared to the latter.

In theory, impact frequency F and F′, as well as impact energy E and E′, should be equal. Naturally, numerical calculations involve errors. However, in the table, the errors in the PUA method are much smaller compared to those in the R-K method. With the R-K method, only when the step size is less than 0.1 ms do the values approach each other; beyond that, the results become almost unreliable. The maximum difference in frequency is 8.64 Hz, and the maximum difference in impact energy is also 14.94 J.

(2) The PUA Method is More stable Than the R-K Method.

It can be seen from Tables 6.1 and 6.2 that as the calculation step size increases, the simulation results from the PUA method exhibit minimal fluctuations and follow a consistent pattern. In contrast, the results from R-K method shows larger fluctuations with less regularity, particularly in terms of the accumulator volume, which is used to assess the stability of the simulation results. The accumulator volume fluctuates significantly with the R-K method, while the PUA method only shows variations of a few tenths of a milliliter.

In conclusion, the PUA method, when used in conjunction with correction calculations, exhibits higher accuracy and better stability compared to the R-K method. The main reason for its high accuracy lies in the setup of the correction procedure. It is evident that the credibility of simulations with uncorrected methods is quite low when applied to more state transitions (ideal models only have three) and simulations with more complex calculation formulas.

6.4 Basic Module of Simulation Computation

Through the extensive work carried out earlier, mathematical models for the various motion states of the hydraulic impact mechanism have been established, and a suitable quasi-constant acceleration numerical calculation method has been proposed. The remaining key task is to convert these sets of equations into expressions suitable for computer iterative calculations, thereby establishing a simulation computation model.

The hydraulic impact mechanism has multiple working states within one operating cycle, making the mathematical models that describe these states relatively complex. To facilitate programming and ease of use, a modular approach has been adopted to establish the simulation model for the impact mechanism. This involves creating common basic simulation computation modules, which can be called during various state simulations. During initialization, it is only necessary to assign values to the public variables in the common model group. Here are two points to note:

- The computational step size is represented by τ, which denotes a time interval.
- Quantities with the superscript '^' represent the values from the previous computational step, while quantities without the superscript represent the values from the current computational step.

The modules are listed according to the sequence of the simulation computation.

1. **Piston Motion Module ($\overline{P}$):**

$$F_p = A_2\hat{p}_2 - A_1\hat{p}_1,$$

$$F_{sp} = sgn(\hat{u}_p)\{B_1\hat{u}_p - sgn(\hat{u}_p)[B_2(\hat{p}_2 - \hat{p}_l) + B_3(\hat{p}_1 - \hat{p}_l)]\},$$

$$F_{lp} = B_4(\hat{p}_2 - \hat{p}_l) + B_5(\hat{p}_1 - \hat{p}_l),$$

$$\omega_p = (F_p - F_{sp} - F_{lp})/m_p,$$

$$\Delta y_p = \hat{u}_p\tau + \frac{1}{2}\omega_p\tau^2,$$

$$y_p = \hat{y}_p + \Delta y_p,$$

$$u_p = \hat{u}_p + \omega_p\tau,$$

$$t_p = \hat{t}_p + \tau,$$

where, t_P: is the time of piston movement;
Δy_p: is the displacement increment of piston (within one step).

2. **Valve Spool Movement Module ($\overline{V}$):**

$$F_v = |A_3(\hat{p} - \hat{p}_l) - A_5(\hat{p} - \hat{p}_5)|,$$

$$F_{sv} = B_6(u_v) - B_7(\hat{p} - \hat{p}_l),$$

$$F_{lv} = B_8(\hat{p} - \hat{p}_l) + B_9(\hat{p} - \hat{p}_5),$$

$$\omega_v = (F_v - F_{sv} - F_{lv})/m_v,$$

$$\Delta y_v = \hat{u}_v\tau + \frac{1}{2}\omega_v\tau^2,$$

$$y_v = \hat{y}_v + \Delta y_v,$$

$$u_v = \hat{u}_v + \omega_v\tau,$$

$$t_v = \hat{t}_v + \tau,$$

where t_v: is the time of spool movement;

$\Delta \boldsymbol{y}_v$: is the displacement increment of the spool within one step.

The absolute value is taken on the right side of the first equation because there is no need to distinguish between positive and negative directions for the displacement y_v during the valve's return stroke and impact stroke reversals. Therefore, the force also does not differentiate between positive and negative values.

3. **Movement Module of Oil Column in Return Pipe ($\overline{O}$):**

$$F_0 = A_0\hat{p}_l,$$

$$F_{s0} = K_{10}\hat{u}_0^2,$$

$$\omega_0 = (F_0 - F_{so})/m_0,$$

$$\Delta y_0 = \hat{u}_0\tau + \frac{1}{2}\omega_0\tau^2,$$

$$y_0 = \hat{y}_0 + \Delta y_0,$$

$$u_0 = \hat{u}_0 + \omega_0\tau,$$

$$t_0 = \hat{t}_0 + \tau.$$

In the equation:

t_0: time of the oil column movement in the return pipe;

Δy_0: displacement increment of the oil column within one time step.

4. **Calculation Module for High-Pressure Flow Rate, Volume and Pressure of High-Pressure Accumulator ($\overline{Q_h}$):**

In Chap. 5, the balance equations for high-pressure flow rate in each state were presented as Eqs. (5.4) and (5.5), etc. However, these are not yet suitable for simulation calculations. Especially in the simulation calculation formulas derived from these equations, the flow rates Q_e and Q_a (the charging and discharging flow rates of the accumulator) are typically calculated first, followed by the calculation of pressure p_h. This approach often leads to flow rate divergence during frequent state transitions, causing the simulation to lose control and ultimately be interrupted. Therefore, a proper transformation is required to obtain the calculation formulas for pressure differential first, and then for the accumulator charging and discharging flow rate Q_a and compensating flow rate Q_e.

The high-pressure flow balance equation can be expressed uniformly as:

$$Q_i = Q_{pv} + Q_a + Q_e + Q_l, \tag{6.16}$$

in the equation, Q_{pv} is the flow rate required when the piston and valve move.

By differentiating Eq. (5.13) with respect to time, we obtain:

$$Q_a = -\frac{dV_h}{dt} = \frac{C_h}{1.4V_h^{0.4}p_h^2} \cdot \frac{dp_h}{dt} = K_a\frac{dp_h}{dt}. \tag{6.17}$$

For easier calculation, the input pressure p of the impact mechanism in the expression for the compensation flow Q_e (Eq. 5.3) is replaced with the pressure p_h of the high-pressure accumulator, yielding:

$$Q_e = \frac{V_k}{K_E} \cdot \frac{dp_h}{dt} = K_v\frac{dp_h}{dt}. \tag{6.18}$$

When the pressure p_h increases, Q_a and Q_e actually store oil, and when the pressure decreases, oil is supplied to the piston and valve. The selection of symbols in the above equation is determined based on this physical meaning.

Substituting Eqs. (6.17) and (6.18) into Eq. (6.16) yields:

$$\frac{dp_h}{dt} = \frac{Q_i - Q_{pv} - Q_l}{K_a + K_v}. \tag{6.19}$$

Based on the above analysis, the following simulation calculation formulas can be written as:

$$K_a = \frac{C_h}{1.4\hat{v}_h^{0.4}\hat{p}_h^2},$$

$$K_v = \frac{V_k}{K_E},$$

$$Q_{pv} = \frac{A_p\Delta y_p + A_v\Delta y_v}{\tau},$$

$$p_h = \hat{p}_h + \frac{(Q_i - Q_{pv} - Q_l)\tau}{K_a + K_v},$$

$$V_h = (C_h/p_h)^{1/1.4},$$

$$V_k = V_{k0}\left(1 + \frac{p_h}{K_e}\right),$$

$$Q_a = K_a \frac{\hat{p}_h - p_h}{\tau},$$

$$Q_e = K_v \frac{\hat{p}_h - p_h}{\tau}.$$

The above eight formulas constitute the simulation computation module ($\overline{Q}_h$). In these formulas, A_p and A_v represent the formal variables of the effective area of high-pressure oil acting on the piston and valve, respectively. They also have state recognition functions and are assigned values during initialization in each state. During the return acceleration phase, $A_p = A_2$; during the return deceleration and impact stroke phase, $A_p = A_2 - A_1$; for the valve during the return stroke reversal phase, $A_v = A_3$, and during the impact stroke reversal phase $A_v = A_5 - A_3$; when the piston and valve are stationary, $A_p = 0$,$A_v = 0$. V_{k0} represents the initial volume of oil in the high-pressure inlet hose, considering the volume when there is no pressure.

5. **Return Oil Accumulator Module ($\overline{C_l}$):**

$$V_l = \hat{V}_l + A_0 \Delta y_0 - A_{11} y_p - \hat{Q}_l \tau,$$

$$p_l = C_l / V_l^{1.4}.$$

In the formula, A_{11} is the formal variable representing the piston's oil discharge area. During the return acceleration phase, oil is discharged, $A_{11} = A_1$. In other phases no oil is discharged, $A_{11} = 0$.

6. **Pressure Differential Calculation Module for the Return Acceleration Phase ($\Delta\overline{p}_r$):**

$$Q_1 = A_1 u_p,$$

$$Q_2 = A_2 u_p,$$

$$Q_3 = Q_i,$$

$$Q_4 = Q_3 - Q_2,$$

$$Q_5 = A_5 u_v,$$

$$p_2 = p_h + sgn(Q_4) K_4 Q_4^2 - K_2 Q_2^2,$$

$$p = p_2 + K_2 Q_2^2 + K_3 Q_3^2,$$

$$p_1 = p_l + K_6 Q_1^2 + x_2 C_8 \left(\frac{Q_1}{S_{2v} - y_v} \right)^2,$$

$$p_3 = x_1 p_l + x_2 (p_2 - K_5 Q_5^2).$$

In the formulas, state identification variables x_1 and x_2 are introduced. When the valve spool is not moving, $x_1 = 1$ and $x_2 = 0$; when spool is moving, $x_1 = 0$ and $x_2 = 1$.

7. **Pressure Differential Calculation Module for Return Braking and Impact Acceleration Phase ($\Delta \overline{p_i}$):**

$$Q_1 = A_1 u_p,$$

$$Q_2 = A_2 u_p,$$

$$Q_3 = Q_i + Q_1 - x_3 Q_{11},$$

$$Q_4 = Q_i + Q_1 - Q_2 - x_3 Q_{11},$$

$$p_2 = p_h + sgn(Q_4) K_4 Q_4^2 - sgn(Q_2) K_2 Q_2^2,$$

$$p = p_2 + sgn(Q_2) K_2 Q_2^2 + sgn(Q_3) K_3 Q_3^2,$$

$$p_1 = x_4 \left\{ p + sgn(Q_1) K_1 Q_1^2 + x_2 sgn(u_p) C_8 \left[Q_1 / (S_{1v} - y_v) \right]^2 \right\}.$$

In the formulas, x_2, x_3 and x_4 are state identification variables. When the valve spool is not moving, $x_2 = 0$, and when it is moving, $x_2 = 1$, When entering state C (positive opening interval), $x_3 = 1$ and $x_4 = 0$; otherwise, $x_3 = 0$ and $x_4 = 1$.

8. **Dedicated Module for Leakage Flow Rate and Compensation Flow Rate ($\overline{Q_l}$)**

$$Q_{l1} = R_1 (p_1 - p_l) + sgn(u_p) R_2 u_p,$$

$$Q_{l2} = R_3 (p_2 - p_l) + sgn(u_p) R_4 u_p,$$

$$Q_{lv} = R_5 (p - p_l),$$

$$Q_l = Q_{l1} + Q_{l2} + Q_{lv}.$$

In addition, the pressure calculation module for the valve spool in the positive opening interval ($\overline{\mathbf{Z}}_0$) and other specialized simulation calculation modules are described in the following section.

6.5 Energy Balance Calculation in Simulation

In addition to impact energy and impact frequency, the energy utilization efficiency is also one of the performance indicators of the hydraulic impact mechanism. In the previous chapters of this book, the efficiency of the impact mechanism was explored with linear models. Although many simplifications were made, the calculation of efficiency remains a complex issue, as it can be considered a function of all working parameters and structural parameters of the impact mechanism. Clearly, the calculation of efficiency, that is the energy balance calculation must be proposed in simulation calculations to better match real operating conditions.

The energy balance equation of the impact mechanism in one working cycle can be expressed as:

$$E_t = E_i + E_v + E_c + E_R + E_k + E_0 + E_l \tag{6.20}$$

(1) E_t: the total input energy, which is the integral of the pressure and flow rate at the control valve port.

$$E_t = \int_0^T pQdt \tag{6.21}$$

(2) E_i: the useful output energy, which is the impact energy of the impact piston.

$$E_i = \frac{1}{2}m_p u_{im}^2 \tag{6.22}$$

(3) E_v: the energy consumed by the valve spool during reversal.

$$E_v = \int_0^{sv} F_{vr}dy_v + \int_0^{sv} F_{vi}dy_v \tag{6.23}$$

where, F_{vr} and F_{vi}, represent the forces driving the valve during return stroke reversal and impact stroke reversal, respectively.

(4) E_c: the energy loss caused by local resistance in the internal oil passages of the impact mechanism. According to the previous calculation modules of $\Delta\overline{P}_r$ and $\Delta\overline{P}_i$, we have:

$$E_c = \int_0^T \left[x_5 K_1 + \frac{C_8}{(y_v - S_{2v})^2} \right] |Q_1^3| dt$$
$$+ \int_0^T \left(K_2|Q_2^3| + K_3|Q_3^3| + K_4|Q_4^3| + x_2 K_5|Q_5^3| \right) dt. \quad (6.24)$$

where the values of the state identifiers x_2, x_5 are:

- when the valve doses not move, $x_2 = 0$;
- when the valve moves: $x_2 = 1$;
- during return stroke acceleration phrase, $x_5 = 0$;
- during return stroke braking and impact stroke acceleration phrase, $x_5 = 1$.

(5) E_R: the energy loss caused by piston viscous resistance.

$$E_R = \int_0^{S_p} F_{sp} dy_p \quad (6.25)$$

(6) E_k: the energy loss caused by hydraulic clamping of the piston.

$$E_k = \int_0^{S_p} F_{lp} dy_p \quad (6.26)$$

(7) E_0: the energy loss caused by return oil resistance.

$$E_0 = \int_0^{S_r} A_1 p_l dy_p. \quad (6.27)$$

(8) E_l: the energy loss caused by leakage

It includes oil leakage losses from the front and rear chambers of the piston, as well as from the valve's chamber.

$$E_l = \int_0^T \left[(p_1 - p_l)Q_{l1} + (p_2 - p_l)Q_{l2} + (p - p_l)Q_{lv}\right]dt. \tag{6.28}$$

Based on the above analysis, it is evident that the calculation of energy losses in hydraulic impact mechanisms is quite complex. Some calculations involve the integration of pressure multiplied by flow rate over time, while others require the integration of force over distance. However, the integrand functions do not have simple analytical expressions and can only be calculated using numerical integration methods. Similar to the previously mentioned quasi-constant acceleration numerical method, where the acceleration within a calculation step τ is treated as constant, the pressure, flow rate, and force within a step τ can also be treated as constants. This leads to the following numerical simulation expressions for energy balance.

$$E_t = \widehat{E}_t + pQ_i\tau,$$

$$E_v = \widehat{E}_v + x_2 F_v \Delta y_v,$$

$$E_c = \widehat{E}_c + \left[x_5 K_1 + \frac{C_8}{(y_v - S_{2v})^2}\right]\left|Q_1^3\right|\tau + \left(K_2\left|Q_2^3\right| + K_3\left|Q_3^3\right| + K_4\left|Q_4^3\right| + x_2 K_5\left|Q_5^3\right|\right)\tau,$$

$$E_R = \widehat{E}_R + F_{Rp}\left|\Delta y_p\right|,$$

$$E_k = \widehat{E}_k + F_{lp}\left|\Delta y_p\right| + x_2 F_{lv}\Delta y_v,$$

$$E_0 = \widehat{E}_0 + x_6 A_1 p_l \Delta y_p,$$

$$E_l = \widehat{E}_l + \left[(p_1 - p_l)Q_{l1} + (p_2 - p_l)Q_{l2} + (p - p_l)Q_{lv}\right]\tau.$$

The above eight equations form the simulation calculation model of the energy balance ($\overline{E}$).

Additionally, the efficiency expressions for the impact mechanism can be derived as follows:

$$\text{total efficiency: } \eta = \frac{E_i}{E_t}$$

$$\text{mechanical efficiency: } \eta_m = \left(1 - \frac{E_R + E_k + E_0}{E_t}\right)$$

$$\text{volumetric efficiency: } \eta_v = \frac{(A_1 - A_2)S_p}{Q_i T}$$

$$\text{pressure efficiency: } \eta_p = \left(1 - \frac{E_c}{E_t}\right)$$

6.6 Flowchart and Description of Simulation Computation

Based on the characteristics of hydraulic impact mechanism simulation, the quasi-constant acceleration numerical method and a correction calculation scheme have been proposed, and eight sets of iterative equations for simulation calculations have been established. The next main task is to develop the simulation program. For such a simulation program with complex state transitions, it is crucial to first develop a detailed and logical program flowchart.

Figures 6.3 and 6.4 present the computer simulation program flowcharts for the rear-control type hydraulic impact mechanism. Further explanations based on the content of the flowcharts are provided below.

1. **Overall Initialization**

The overall initialization includes reading data, calculating various coefficients and parameters, setting flags, and inputting other necessary information.

The main idea of this simulation program is to input a certain flow rate to a hydraulic impact mechanism with predetermined structural parameters and let it "operate" until it reaches a stable state. Then, we examine whether its performance parameters (impact energy, impact frequency, efficiency) and working parameters (working oil pressure, stroke, etc.) meet the requirements.

The structural parameters and system parameters of the specific hydraulic impact mechanism are compiled into an Excel file, which the program reads to initialize the structure of the hydraulic impact mechanism.

The expressions for the coefficients that are more complex to compute in the simulation model are summarized below:

$$B_0 = \frac{\pi\mu}{\sqrt{1-\varepsilon^2}}; \qquad B_1 = B_0\left(\frac{dl}{h} + \frac{dl_6}{h} + \frac{d_1 l_1}{h_1} + \frac{d_2 l_2}{h_2}\right);$$

$$B_2 = \frac{\pi}{2}(d_2 h_2 - dh); \qquad B_3 = \frac{\pi}{2}(dh - d_1 h_1);$$

$$B_4 = \tau(dl + d_2 l_2); \qquad B_5 = \tau(dl_6 + d_1 l_1);$$

$$B_6 = B_0\left(\frac{d_3 l_3}{h_3} + \frac{2d_4 l_4}{h_4} + \frac{d_5 l_5}{h_5}\right); \qquad B_7 = \frac{\pi}{2} d_3 h_3;$$

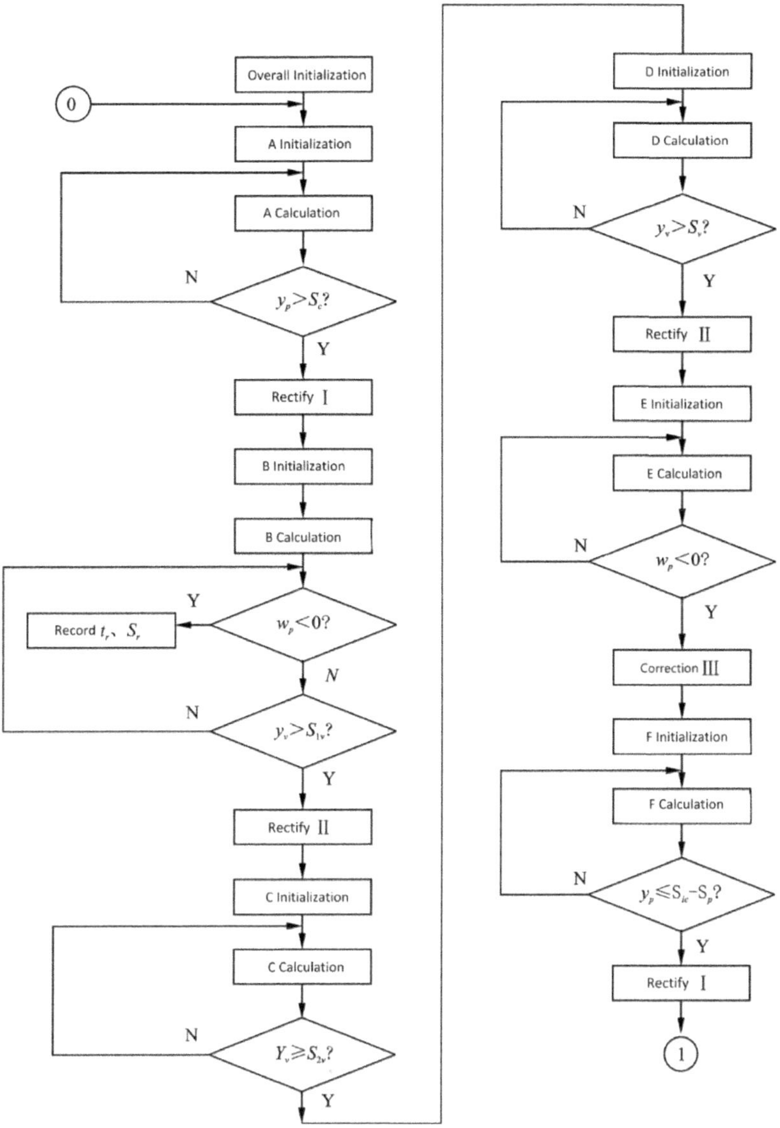

Fig. 6.3 Flowchart of the simulation program for the rear-control hydraulic impact mechanism

$$B_8 = \tau(d_3l_3 + 2d_4l_4); \qquad B_9 = \tau d_5l_5;$$

$$R_0 = \frac{\pi(1 + 1.5\varepsilon^2)}{12\mu}; \qquad R_1 = R_0\left(\frac{dh^3}{l_6} + \frac{d_1h_1^3}{l_1}\right);$$

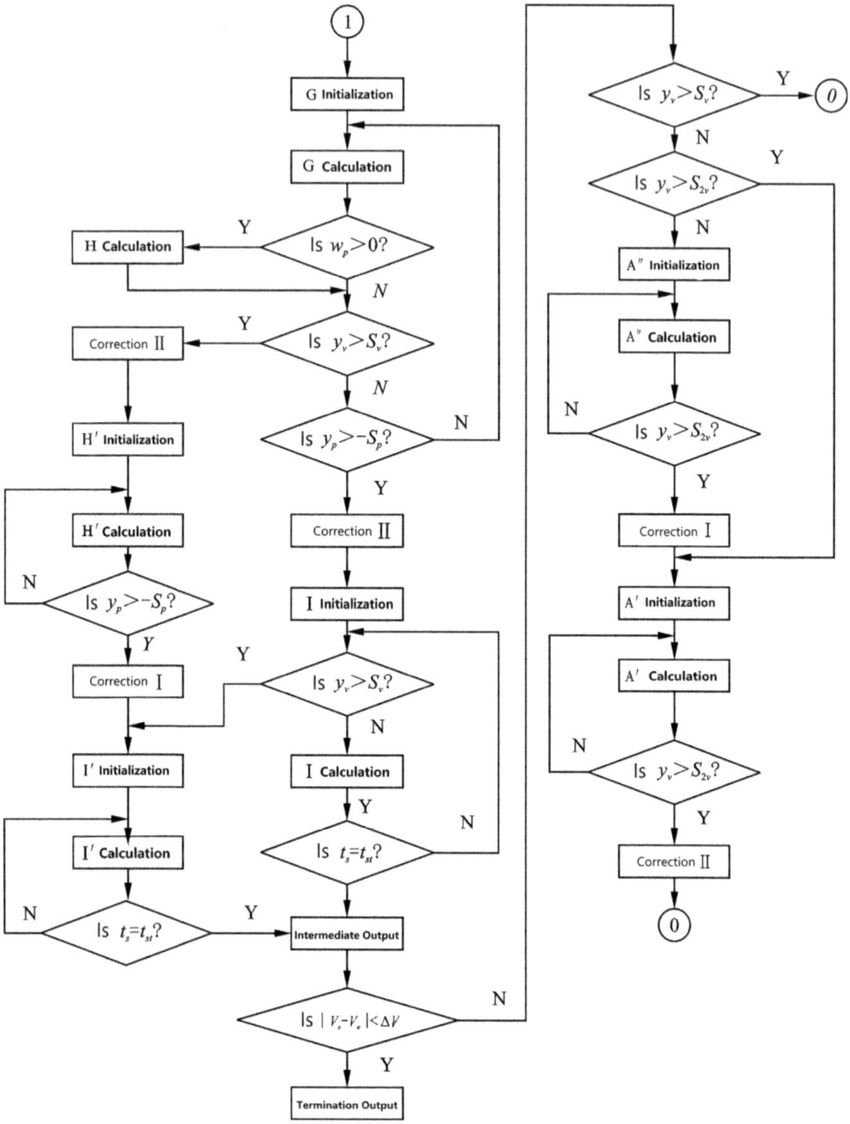

Fig. 6.4 Flowchart of the simulation program for the rear control hydraulic impact mechanism

$$R_2 = \frac{\pi}{2}(d_1h_1 - dh); \qquad R_3 = R_0\left(\frac{dh^3}{l} + \frac{d_2h_2^3}{l_2}\right);$$

$$R_4 = \frac{\pi}{2}(dh - d_2h_2); \qquad R_5 = R_0\left(\frac{d_3h_3^3}{l_3} + \frac{2d_4h_4^3}{l_4} + \frac{d_5h_5^3}{l_5}\right);$$

$$K_1 = \frac{\zeta_1 \rho}{2A_8^2} : K_2 = \frac{\zeta_2 \rho}{2A_8^2}; \qquad K_3 = \frac{\zeta_3 \rho}{2A_8^2} : K_4 = \frac{\zeta_4 \rho}{2A_8^2}; K_5 = \frac{\zeta_5 \rho}{2A_7^2} : K_6 = \frac{\zeta_6 \rho}{2A_8^2};$$

$$C_7 = C\pi d_4 \sqrt{2/\rho} : C_8 = \frac{\rho}{2C^2\pi^2 d_4^2} = \frac{1}{C_7^2}.$$

In addition, the determination of several coefficients for calculating the compensating flow rate Q_e caused by oil compression and hose expansion is relatively complicated.

The part of the compensation flow rate comes from the volume change of the internal oil in the mechanism, denoted as V_2, while another part comes from the volume change of the high-pressure hose for external oil intake, denoted as V_1. The sum of the volumes of these two parts is V_k. Due to the fact that the internal oil in the impact mechanism is surrounded by a body with relatively significant stiffness, the deformation of the body caused by pressure changes can be neglected. Thus, the total volume increment caused by the increase in pressure can be expressed as:

$$\Delta V_k = \Delta V_{1c} - \Delta V_{1l} - \Delta V_{2l}.$$

In the equation:

hose expansion increment

$$\Delta V_{1c} = V_1 \cdot \frac{\Delta p}{K_c}$$

Increment of oil compression in the hose.

$$\Delta V_{1l} = -V_1 \cdot \frac{\Delta p}{K_l}$$

Increment of oil compression in the body.

$$\Delta V_{2l} = -V_2 \cdot \frac{\Delta p}{K_l}$$

where, K_c and K_l represent the volumetric elastic moduli of the hose and the oil, respectively. Introducing the apparent elastic modulus K_E, the total volume increment can be written as:

$$\Delta V_k = (V_1 + V_2) \cdot \frac{\Delta p}{K_E} = V_k \frac{\Delta p}{K_E}.$$

Assuming the oil pressure inside the mechanism and the pressure in the high-pressure inlet hose are uniform throughout, the following can be derived from the above equations:

$$\frac{V_k}{K_E} = \frac{V_1}{K_c} + \frac{V_1}{K_l} + \frac{V_2}{K_l},$$

$$K_E = \frac{V_k K_c K_l}{V_k K_c + V_1 K_l},$$

$$K_v = \frac{V_k}{K_E}.$$

The last two equations are the calculation formulas for the two coefficients used in the compensation flow rate calculation.

In the simulation calculations, sometimes it's necessary to distinguish the compensating flow rate generated by V_2 from the total compensation flow rate. To achieve this, an additional calculation coefficient is introduced,

$$K_{2k} = \frac{\Delta V_{2l}}{\Delta V_k} = \frac{V_2 K_c}{V_1 K_c + V_1 K_l + V_2 K_c} = \frac{V_2 K_c}{V_k K_c + V_1 K_l},$$

$$K_E = \frac{K_c K_l}{K_c + K_l}, \ K_v = \frac{V_k}{K_E}.$$

2. **Initialization and Simulation Calculation Model of Each State**

In Sect. 6.4, some basic calculation modules for simulation were established. Next, based on the flowchart, the simulation model for each working state will be determined, including the composition of the basic modules and the values of state recognition variables in the initialization calculation module. For the sake of simplicity, these are summarized in Table 6.3.

6.7 Simulation Computation Under Special Conditions

When the hydraulic impact mechanism goes through states A″ and C, although the time duration is short, the calculations involved are relatively complex and require some processing to facilitate programming. Additionally, when the high-pressure accumulator is not in operation, the oil pressure and flow rate characteristics of the impact mechanism will differ significantly, and the relevant calculation formulas must be adjusted accordingly.

1. **State A″**

The abnormal working state caused by a significant delay in the impact stroke reversal has been described earlier. After the piston completes its impact, if the valve spool displacement $y_v < S_{1v}$ (i.e., it has not yet entered the positive opening interval), the piston will decelerate with its initial velocity gained from the rebound, and secondary impacts may even occur. If the spool is within the positive opening interval ($S_{1v} \leq$

Table 6.3 Summary of initialization of each state mathematical model (Code)

"A calculation"	$(\overline{P})$,$(\overline{O})$, $(\overline{Q}_h)$, $(\overline{C}_l)$, $(\overline{\Delta P_r})$, $(\overline{Q}_l)$, $(\overline{E})$
"A initialization"	$A_P = A_2, A_v = 0, A_{11} = A_1, x_1 = 1, x_2 = 0, x_3 = 0, x_4 = 1, x_5 = 0, x_6 = 1$
"B calculation"	$(\overline{P})$, $(\overline{V})$, $(\overline{Q}_h)$, $(\overline{C}_l)$, $(\overline{O})$, $(\overline{\Delta P_r})$, $(\overline{Q}_l)$, $(\overline{E})$
"B initialization"	$A_P = A_2, A_v = A_3, A_{11} = A_1, x_1 = 0, x_2 = 1, x_3 = 0, x_4 = 1, x_5 = 0,$ $x_6 = 1$
"C calculation"	$(\overline{P})$, $(\overline{V})$, $(\overline{Q}_h)$, $(\overline{C}_l)$, $(\overline{O})$, $(\overline{Z}_0)$, $(\overline{\Delta P_i})$, $(\overline{Q}_l)$, $(\overline{E})$
"C initialization"	$A_P = A_2 - A_1, A_v = A_3, A_{11} = 0, x_1 = 0, x_2 = 1, x_3 = 1, x_4 = 0, x_5 = 0,$ $x_6 = 1$
"D calculation"	$(\overline{P})$, $(\overline{V})$, $(\overline{Q}_h)$, $(\overline{C}_l)$, $(\overline{O})$, $(\overline{\Delta P_i})$, $(\overline{Q}_l)$, $(\overline{E})$
"D initialization"	$A_P = A_2 - A_1, A_v = A_3, A_{11} = 0, x_1 = 0, x_2 = 1, x_3 = 0, x_4 = 1, x_5 = 0,$ $x_6 = 1$
"E calculation"	$(\overline{P})$, $(\overline{Q}_h)$, $(\overline{C}_l)$, $(\overline{O})$, $(\overline{\Delta P_i})$, $(\overline{Q}_l)$, $(\overline{E})$
"E initialization"	$A_P = A_2 - A_1, A_v = 0, A_{11} = 0, x_1 = 1, x_2 = 0, x_3 = 0, x_4 = 1, x_5 = 0,$ $x_6 = 1$
"F calculation" "F initialization"	Same as E status
"G calculation"	$(\overline{P})$, $(\overline{V})$, $(\overline{Q}_h)$, $(\overline{C}_l)$, $(\overline{O})$, $(\overline{\Delta P_i})$, $(\overline{Q}_l)$, $(\overline{E})$
"G initialization"	$A_P = A_2 - A_1, A_v = A_5 - A_3, A_{11} = 0, x_1 = 0, x_2 = 1, x_3 = 0,$ $x_4 = 1, x_5 = 0, x_6 = 1$
"H calculation"	$(\overline{P})$, $(\overline{V})$, $(\overline{Q}_h)$, $(\overline{C}_l)$, $(\overline{O})$, $(\overline{\Delta P_i})$, $(\overline{Q}_l)$, $(\overline{E})$
"H initialization"	$A_P = A_2, A_v = A_5 - A_3, A_{11} = 0, x_1 = 0, x_2 = 1, x_3 = 0, x_4 = 1, x_5 = 0,$ $x_6 = 1$
"I calculation"	$(\overline{V})$, $(\overline{Q}_h)$, $(\overline{C}_l)$, $(\overline{O})$, $(\overline{Q}_l)$, $(\overline{E})$
"I initialization"	$A_P = 0, A_v = A_5 - A_3, A_{11} = 0, x_1 = 0, x_2 = 1, x_3 = 0, x_4 = 1, x_5 = 0,$ $x_6 = 1$
"I′ calculation"	$(\overline{Q}_h)$, $(\overline{C}_l)$, $(\overline{O})$, $(\overline{Q}_l)$, $(\overline{E})$
"I′ initialization"	$A_P = 0, A_v = 0, x_2 = 0, x_3 = 0, x_4 = 1, x_5 = 0, x_6 = 1$
"A′ calculation"	$(\overline{P})$, $(\overline{V})$, $(\overline{Q}_h)$, $(\overline{C}_l)$, $(\overline{O})$, $(\overline{\Delta P_r})$, $(\overline{Q}_l)$, $(\overline{E})$
"A′ initialization"	$A_P = A_2, A_v = A_5 - A_3, A_{11} = A_1, x_1 = 0, x_2 = 1, x_3 = 0, x_4 = 1, x_5 = 0,$ $x_6 = 1$
"H′ calculation"	$(\overline{P})$, $(\overline{V})$, $(\overline{Q}_h)$, $(\overline{C}_l)$, $(\overline{O})$, $(\overline{\Delta P_r})$, $(\overline{Q}_l)$, $(\overline{E})$
"H′ initialization"	$A_P = A_2, A_v = A_5 - A_3, A_{11} = 0, x_1 = 0, x_2 = 1, x_3 = 0, x_4 = 1, x_5 = 0,$ $x_6 = 1$

$y_v \leq S_{2v}$) at this moment, the rear chamber of the piston will be almost in a closed state, and the rebound speed of the piston will be almost entirely suppressed due to the constraints of the rear chamber. In either of the above cases, it will result in prolonging the piston impact dwell time and wastefully dissipating most of the rebound energy. Based on this characteristic, when this state is detected during the simulation process, it is approximately treated with the following method:

- Set the piston rebound speed to zero.

- Continue the simulation computation as in the dwell state I until $y_v \geq S_{2v}$.
- Output the delayed stroke and the delayed time of the valve spool, and indicate the occurrence of the abnormal state A″.

2. **State C**

This is a brief state where the valve is in the positive opening interval. However, significant pressure changes occur during this period, and if calculated with the regular (time) step size, only a very few steps or even less than one step would "jump" over, making it impossible to reflect the actual situation of this state. Therefore, a relatively special method must be used to calculate the pressure changes within the positive opening interval.

(1) Simplify the Original Calculation Formula

As the return oil pressure p_l is much lower compared to p and p_1, its influence becomes even smaller during an approximate transient process. Therefore, by setting $p_l = 0$, the original Eqs. (4.29)–(4.31) can be simplified as:

When $p_1 > p$,

$$p_1 = \frac{2ab - p - \sqrt{p^2 - 4abp + 4b^2}}{2(a^2 - 1)}. \tag{6.29}$$

When $p_1 < p$,

$$p_1 = \frac{2cd + p + \sqrt{p^2 + 4cdp - 4d^2}}{2(c^2 + 1)}. \tag{6.30}$$

When $z_1 = z_2$,

$$p_1 = \frac{b^2}{2b - p}. \tag{6.31}$$

where:

$$a = \frac{z_1^2 + z_2^2}{2z_1 z_2},$$

$$b = \frac{C_8 A_1^2 u_p^2 + z_2^2 p}{2z_1 z_2},$$

$$c = \frac{z_1^2 - z_2^2}{2z_1 z_2},$$

$$d = \frac{C_8 A_1^2 u_p^2 - z_2^2 p}{2z_1 z_2}.$$

(2) How to Determine the Relationship Between p_1 and p?

If Eqs. (6.29) and (6.30) are unified and written as:

$$p_1 = \frac{2ef - ip - i\sqrt{p^2 - 4iefp + 4f^2}}{2(e^2 - i)}, \tag{6.32}$$

where:

$$e = \frac{z_1^2 + iz_2^2}{2z_1z_2}, \tag{6.33}$$

$$f = \frac{C_8A_1^2u_p^2 + iz_2^2p}{2z_1z_2} = \frac{C_8Q_1^2 + iz_2^2p}{2z_1z_2} \tag{6.34}$$

$$i = \pm 1 \tag{6.35}$$

Obviously, when $i = 1, p_1 > p$, and when $i = -1, p_1 < p$. The actual relationship between p_1 and p is determined by the valve openings z_1 and z_2. The calculation judgment is based on the sign of the value inside the square root in Eq. (6.32). When the valve's movement generates a pair of z_1 and z_2, we can first set $i = 1$ and calculate the value inside the square root. If it is positive, it indicates that $p_1 > p$; otherwise, if it is negative, it means $p_1 < p$, and we can set $i = -1$ and perform the calculation again.

Additionally, when $e^2 - i = 0$, it means that $z_1 = z_2$, and Eq. (6.31) should be used to calculate p_1.

(3) Flow Rate Calculation

$$Q_{11} = C_7z_1\sqrt{p_1}, \tag{6.36}$$

$$Q_{12} = sgn(p_1 - p)C_7z_2\sqrt{|p - p_1|}. \tag{6.37}$$

(4) Determination of the Calculation Step Size

The positive opening interval is generally only a few tenths of a millimeter long, and the spool, moving at several meters per second, passes through it in just a few tenths of a millisecond. Therefore, the same simulation step size as in other states is impractical. On the contrary, the step size determined solely on the conditions of the positive opening range would significantly prolong the overall simulation time. Only a displacement step control for this interval is more appropriate.

First, divide the positive opening amount z_0 into n equal parts. To avoid the occurrence of $z_1 = z_2$, it is advisable to divide it into an odd number of parts. Thus, we have:

$$\Delta z = z_0/n,$$

$$z_2' = \hat{z}_2 + \Delta z,$$

$$z_1' = z_0 - z_2'.$$

The valve spool's velocity after passing a displacement of Δz is:

$$u_v'^2 - u_v^2 = 2\omega_v' z,$$

$$u_v' = \sqrt{2\omega_v' \Delta z + u_v^2}.$$

Then the time for the spool to travel through Δz is $\Delta t_v = \frac{u_v' - \hat{u}_v}{\omega_v'}$.

Using the above simulation calculation formulas and Eqs. (6.13)–(6.14), the pressure variation in the rear chamber within the positive opening interval can be calculated.

In this way, in state C, the variation of the rear chamber pressure is calculated with the displacement Δz as the step size, while the valve and piston use the time Δt_v as the step size. However, this step size is variable.

3. **Simulation Calculation Handling When the High-Pressure Accumulator is not in Operation**

Since no high-pressure accumulator is involved in the operation, the 'surplus' or 'deficiency' in flow rate demand during the operation of the impact mechanism can only be compensated by the compression and expansion of the oil in the high-pressure chamber, along with the volume changes in the inlet hose.

When the high-pressure accumulator is not in operation, the $\frac{dV_h}{dt}$ term in the various high-pressure hydraulic oil balance equations becomes zero, i.e., $Q_a = 0$, and Eq. (6.16) becomes:

$$Q_i = Q_{pv} + Q_e + Q_l. \tag{6.38}$$

that is,

$$-\frac{dV_k}{dt} = Q_i - Q_{pv} - Q_l. \tag{6.39}$$

In addition, when the oil pressure increases from zero to p_h, the oil volume increases from V_{k0} to V_k, resulting in the following equation:

$$\Delta V_k = V_k - V_{k0} = V_{k0} \cdot \frac{p_h}{K_E},$$

that is

$$p_h = \left(\frac{V_k}{V_{k0}} - 1\right) K_E. \tag{6.40}$$

This leads to the high-pressure flow balance calculation module ($\overline{C}_h$) when the high-pressure accumulator is not in operation,

$$Q_{pv} = \frac{A_p \Delta y_p + A_v \Delta y_v}{\tau},$$

$$\Delta V_k = (Q_{pv} + Q_l - Q_i)\tau,$$

$$V_k = \widehat{V}_k - \Delta V_k,$$

$$p_h = \left(\frac{V_k}{V_{k0}} - 1\right) K_E,$$

$$Q_e = \frac{\Delta V_k}{\tau},$$

$$Q_{e2} = K_{2k} Q_e.$$

Q_{e2} is the compensating flow rate generated by the internal medium volume V_2 of the impact mechanism.

Of course, the whole simulation program has to be modified accordingly.

6.8 Examples of Simulation Computation

Based on the above operations, a computer simulation program for the hydraulic impact mechanism was developed. Appendix B contains the program listing and its description.

The basic parameters of YYG250 hydraulic impact mechanism are as follows:

Piston mass: $m_p = 12.95$ kg

Spool mass: $m_v = 0.3$ kg (including the mass of oil which moves axially with the spool).

Diameter of the three steps of the piston.

$$d = 0.67\,\text{m} \qquad d_1 = 0.051\,\text{m} \qquad d_2 = 0.062\,\text{m}$$

Diameter of the three steps of the spool.

$$d_3 = 0.014\,\text{m} \qquad d_4 = 0.04\,\text{m} \qquad d_5 = 0.020\,\text{m}$$

Control port position of the spool return stroke reversal $S_c = 0.034$ m
Control port position of the spool impact stroke reversal $S_{ic} = 0.007$ m
Inflation volume of high-pressure accumulator $V_a = 225 \times 2 = 450$ mL
Inflation volume of low-pressure accumulator $V_{la} = 225 \times 2 = 450$ mL
Input flow rate $Q_i = 120.85$ L/min

After inputting the above basic parameters and other relevant parameters into the simulation program shown in Appendix B, the following simulation results can be output.

E_i	391.69	f	30.33				
E_{ta}	0.51	E_{tam}	0.7	E_{tav}	0.7	E_{tap}	0.88
P_{max}	12.07	P_{min}	10.5	P_{ave}	11.49		
u_{rm}	4.32	u_{im}	7.78				
T_r	17.02	T_b	4.04	T_i	11.31		
T_{vr}	2.08	T_{vi}	1.87				
S_p	47.56	S_r	39.09	S_b	8.47		
S_c	34	S_{ic}	7				
Q_i	120.85	P_a	6	P_{la}	0.7	K	0
M_p	12.95	M_v	0.3	V_a	450	V_{la}	450
A_1	14.78	A_2	5.02	A_3	1.53	A_5	3.13

The meanings of the symbols are explained in Appendix B. It is important to note that the overall efficiency E_{ta} is not the product of mechanical efficiency E_{tam}, volumetric efficiency E_{tav}, and pressure efficiency E_{tap}, as these three factors overlap in certain areas. Additionally, the piston areas A_1, A_2, etc., represent the effective areas after subtracting the clearance gaps.

Table 6.4 presents a comparison between the measured results and the simulation results.

Note:

① Pressure refers to the time-average value, calculated as $p_{ave} = \frac{1}{T}\int_0^T pdt$, and T is the working period.
② The error is relative to the measured value.

Table 6.4 Comparison of measured and simulation results

	Energy of impact /J	Frequency of impact /Hz	Velocity of impact /(m · s^{-1})	Maximum return velocity /(m · s^{-1})	Stroke /mm	Pressure /Mpa	Efficiency /%
Experiment	396	28.71	7.82	4.35	48.19	11.9	47.6
Simulation	396.69	30.33	7.78	4.32	47.56	11.49	50.93
Error (%)	−1.0	5.6	−0.5	−0.68	−1.3	−0.34	6.9

From the comparison in Table 6.4, it can be observed that the simulation results obtained from this program show good consistency with the measured results. It is worth noting that many variables in the simulation calculations exhibit significant discretization and randomness. For example, eccentricity, local resistance coefficients at different locations, hydraulic jamming coefficients, etc., can only be appropriately determined with substantial practical experience in the design and research of hydraulic impact mechanisms. Certainly, the measured results of existing models provide an essential basis for adjusting these "soft" coefficients. Without accounting for these factors, the credibility of the simulation results would be considerably low.

Figures 6.5 and 6.6 are plotted based on the simulation results of the above examples, which allow us to have a more visual understanding of the working process of the hydraulic impact mechanism.

This simulation program includes simulations for various normal and abnormal operating conditions. Table 6.5 presents a comparison of simulation results for several operating conditions.

In the table, 'Normal' indicates that the spool's stroke reversal control port position S_{ic} is normal. 'Advanced' means S_{ic} is too large, causing the spool to reverse

Fig. 6.5 Piston velocity u_p, valve velocity u_v (dashed line), inlet Pressure$p$, rear chamber pressure p_1 and oil return pressure p_l

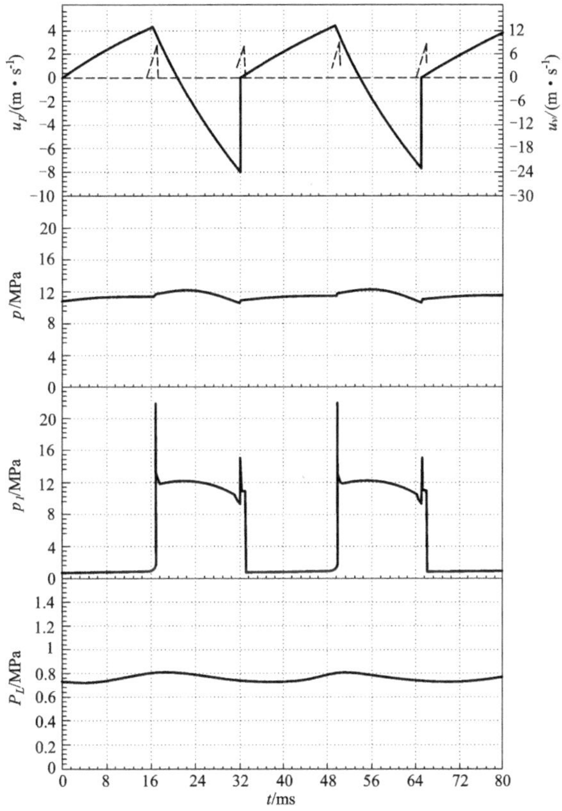

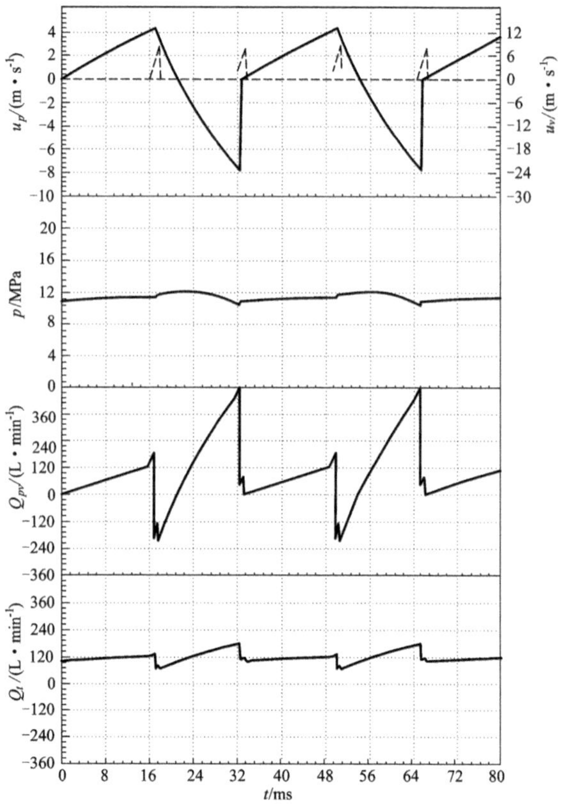

Fig. 6.6 Piston velocity u_p, valve velocity u_v (dashed line), inlet Pressure p, required flow Rate of piston and valve Q_{pv}, and inlet flow Q_t

Table 6.5 Comparison of simulation results of several working conditions

		Energy of impact /J	Frequency of impact /Hz	Velocity of impact /m · s^{-1}	Stroke /mm	Pressure /Mpa	Efficiency /%
Reversal normal	No rebound	391.6	30.33	7.78	47.56	11.49	50.93
	Rebound	310.8	30.89	6.93	47.59	9.18	51.9
Reversal advance	No rebound	363.4	35.95	7.49	48.2	16.62	38.49
	Rebound	299.04	35.16	6.8	48.15	12.58	40.97
Reversal delay	No rebound	408.95	30.25	7.95	47.72	11.78	52.5
	Rebound	400.85	30.44	7.87	47.64	11.65	51.74

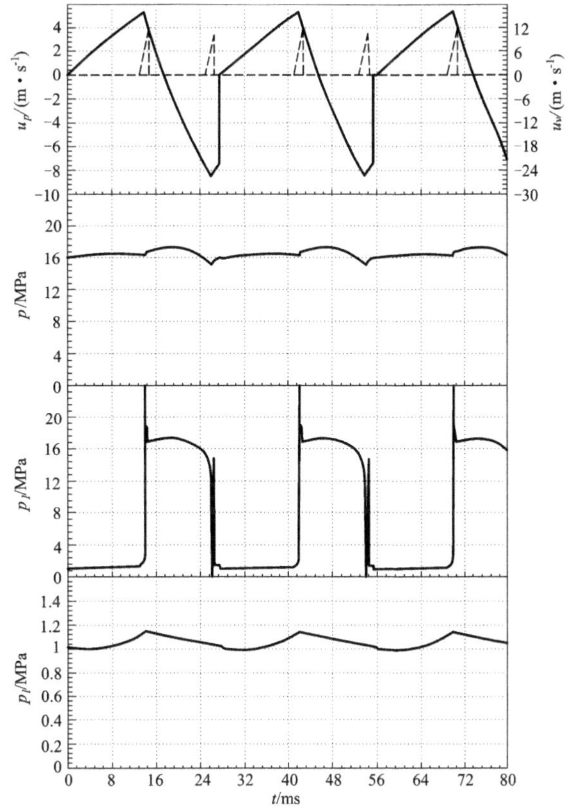

Fig. 6.7 Piston and spool velocities u_p, u_v, inlet pressure$p$, rear chamber pressure p_1, and return oil pressure p_l when the spool impact stroke reversal is advanced

prematurely, with the mechanism entering states H and H′ in the latter stage of the stroke. 'Delayed' means that S_{ic} is too small, resulting in delayed spool reversal, and entering state A″ after the impact pause. Due to state A′ not affecting the piston rebound, it is also considered a normal state.

'Rebound' refers to the phenomenon of the piston having a rebound velocity after impact. When rebound occurs, the rebound coefficient is taken as $k = 0.2$.

Figures 6.7 and 6.8 further illustrate the effects of advanced or delayed spool stroke reversal on pressure when there is no rebound ($k = 0$).

It is worth mentioning that the quasi-constant acceleration numerical method and the correction calculation method for state transitions form the foundation for this program's ability to perform fast calculations with high accuracy. Furthermore, the handling of high-pressure flow balance calculations is crucial for eliminating divergence in the simulation process and achieving stable solutions when considering factors such as the compressibility of the working medium and the expansibility of the high-pressure hose.

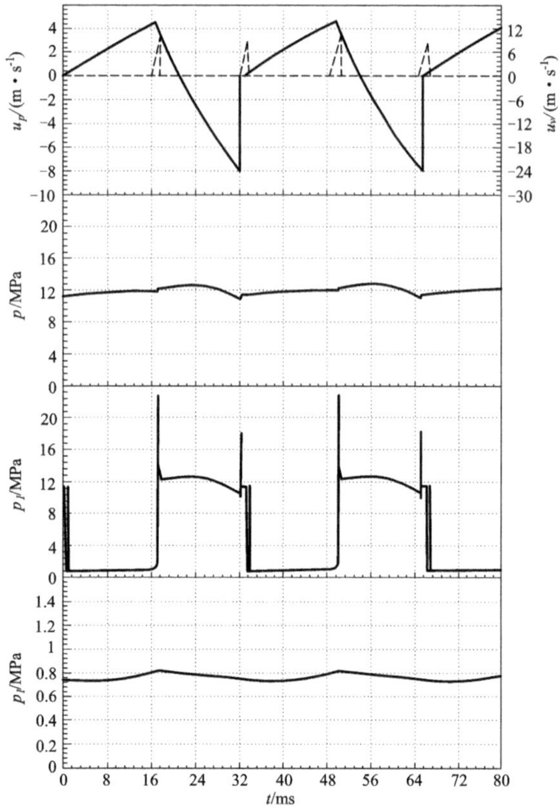

Fig. 6.8 Piston and spool velocities u_p, u_v, inlet pressure$p$, rear chamber pressurep_1, and return oil pressure p_l when the spool impact stroke reversal is delayed

Chapter 7
Several Special Topics on Hydraulic Impact Mechanisms Research

7.1 Role of Compensation Flow in Hydraulic Impact Mechanisms

In the existing research, the role of high-pressure accumulators in recovering return flow, achieving peak oil compensation, and maintaining smooth operating pressure has been well explained. However, there is still insufficient research on whether the flow at the inlet port remains relatively constant, why high-pressure accumulators do not have as significant an effect on the impact energy as expected, and why they have little effect on the impact frequency. In this chapter, we will focus on studying the variation of the flow at the inlet port of the hydraulic impact mechanism and explore these issues in-depth. Earlier sections have dedicated considerable space to describing compensation flows generated by oil compression and hose expansion, which serves as theoretical preparation for the analysis in this chapter.

When establishing the mathematical model of hydraulic impact mechanisms, the compressibility of the fluid and the expansibility of the high-pressure hose are taken into account. The flow related to their apparent elastic modulus is referred to as the compensation flow rate. To what extent can the flow caused by volume changes, which is generally negligible, influence the flow variation in hydraulic impact mechanisms? First, let's look at two measured examples, as shown in Table 7.1.

The oil pressure areas of the front and rear chambers of the YYG90A are $A_2 = 2.14 \times 10^{-4}\,\mathrm{m}^2$ and $A_1 = 7.17 \times 10^{-4}\,\mathrm{m}^2$, respectively. For the YYG220A, the oil pressure areas of the front and rear chambers are $A_2 = 3.77 \times 10^{-4}\,\mathrm{m}^2$ and $A_1 = 11.71 \times 10^{-4}\,\mathrm{m}^2$, respectively. The maximum flow rate in Table 7.1 refers to the flow required to meet the piston's maximum impact velocity u_{im}, and is calculated as $Q_{\max} = u_{im}(A_1 - A_2)$. It does not include the flow rate needed for leakage and valve movement. The "Impact velocity that the pump can meet" u_{im} in Table 7.1 refers to the maximum impact velocity that can be achieved by the pump (using a fixed-displacement pump) with the flow rate Q_i, and is given by $u_{im} = Q_i/(A_1 - A_2)$.

Q. He, *Research and Design of Hydraulic Impact Mechanisms*,
https://doi.org/10.1007/978-981-96-4864-1_7

Table 7.1 Parameters of YYG90A, YYG220A Type Hydraulic Impact Mechanisms

Group		Paramater					
		Frequency /Hz	Impact speed/ $(m \cdot s^{-1})$	Maximum Flow/ $(L \cdot min^{-1})$	Impact speed that the pump can meet/ $(m \cdot s^{-1})$	Average pump flow/ $(L \cdot min^{-1})$	Efficiency/ %
YYG90A	With accumulators	52.2	6.8	205	2.14	64.5	47.5
	Without accumulators	52	6.24	188.2			38.6
YYG220A	With accumulators	31.8	4.75	226.3	1.37	65.2	36.2
	Without accumulators	30.91	4.42	210.6			33

Clearly, this value is greater than the actual maximum impact velocity considering factors like leakage.

From the data in Table 7.1, it can be observed that the impact velocity achieved solely by the pump's output flow rate is much lower than the measured impact velocity. In cases where there is a high-pressure accumulator, the difference in flow rate is naturally provided by the accumulator. But for situations without a high-pressure accumulator, where does this difference in flow rate come from? According to conventional concepts, the impact frequency and impact velocity of hydraulic impact mechanisms would decrease significantly, and there would be a considerable gap in performance indicators from normal conditions. However, this is not the case. Even without a high-pressure accumulator, the impact frequency hardly changes, and the decrease in impact velocity is not much. The only noticeable change is a decrease in efficiency, meaning that more input energy is required. This outcome can only be attributed to the presence of storage devices with accumulator-like functions. In practice, the compressibility of hydraulic oil and the expansion of high-pressure rubber hoses play a role in oil charging and discharging during pressure changes. Although the absolute amount of change is small, for medium and small hydraulic impact mechanisms, the oil consumption per cycle is relatively small, and the compensating effect is still sufficient.

Figures 7.1 and 7.2 show the measured curves of piston velocity, inlet pressure, and flow rate for a hydraulic impact mechanism with and without a high-pressure accumulator, respectively. The data were obtained by a computer-aided testing system for hydraulic impact mechanisms, which was independently designed. Except for the presence or absence of the accumulator, the other working conditions were essentially the same. The graphs also show that the impact frequency and impact velocity are almost the same in both cases. However, there is a significant difference in the degree of variation in the inlet flow rate and pressure curves. The coordinate scales in both

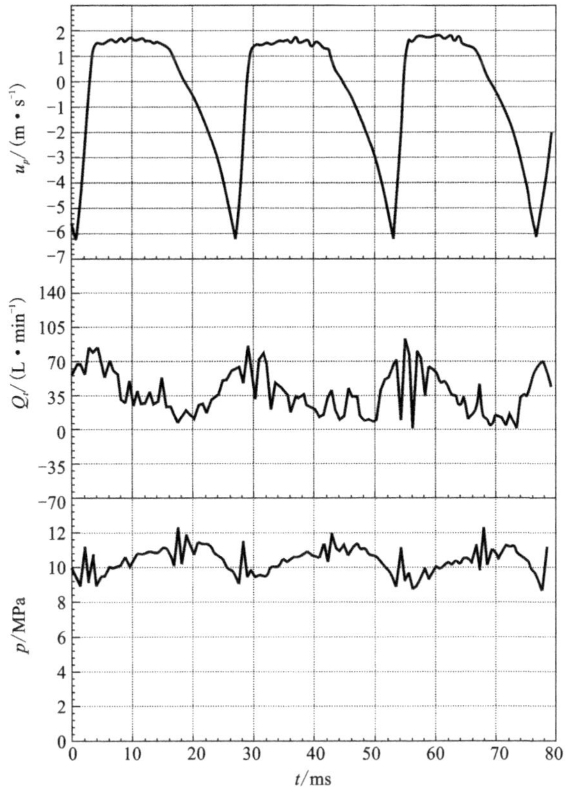

Fig. 7.1 Measured waveform of the piston velocity u_p, inlet flow rate Q_i, and inlet pressure p (with a high-pressure accumulator)

graphs are identical, making the similarities and differences between the two cases are evident at a glance.

If the pump's supply flow rate Q_i, the accumulator flow rate Q_a, the compensation flow rate Q_{e1} generated by the volume change of the oil supply hose, and the compensation flow rate Q_{e2} caused by the volume change of the medium in the internal chamber of the hydraulic impact mechanism are regarded as sources of pressurized oil, while the flow rate Q_{pv} required for piston and valve movement and the leakage flow rate Q_l are regarded as pressurized oil consumption, then we have:

$$Q_i + Q_a + Q_{e1} + Q_{e2} = Q_{pv} + Q_l \tag{7.1}$$

In the equation, except for Q_i and Q_l, the signs are subject to change depending on the operating conditions, with Q_{pv} playing the dominant role. Below, we focus on the case where there is no high-pressure accumulator, i.e., $Q_a = 0$. In this case, Eq. (7.1) becomes:

$$Q_i + Q_{e1} + Q_{e2} = Q_{pv} + Q_l \tag{7.2}$$

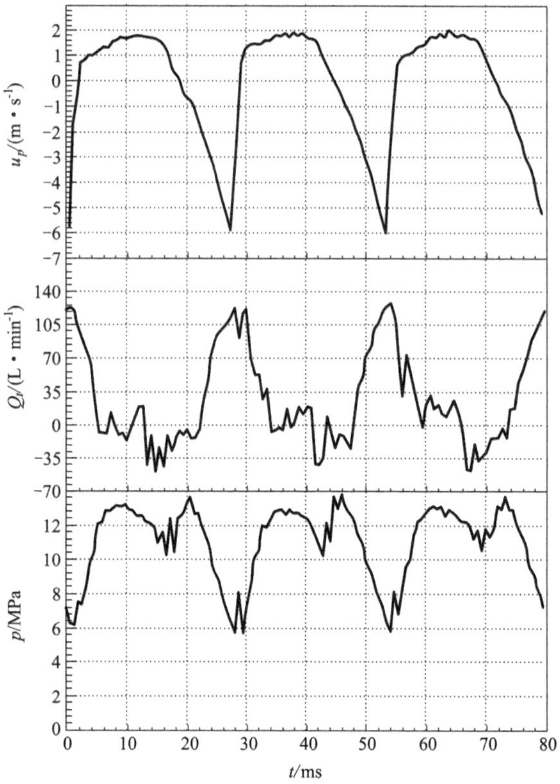

Fig. 7.2 Measured waveform of piston velocity u_p, inlet flow rate Q_t, and inlet pressure p (without high pressure accumulator)

Obviously, the medium flow rate Q_i at the inlet of the hydraulic impact mechanism is not solely the supply flow rate Q_i of the hydraulic pump, as is commonly believed. In reality, it is the combination of the pump flow rate Q_i and the compensation flow rate Q_{e1} generated by the volume change of the high-pressure hose. That is:

$$Q_t = Q_i + Q_{e1} \tag{7.3}$$

The flow rate Q_{pv}, which is related to the movement of the piston and valve (mainly the piston), is the primary factor causing significant fluctuations in the working medium flow in the hydraulic impact mechanism. The sign of Q_{pv} in Eqs. (7.1) and (7.2) can be positive (during the return stroke acceleration and impact stroke acceleration phases) or negative (during the return stroke deceleration phase). The variation curve of Q_{pv} shown in Fig. 6.6 clearly illustrates the degree of its significant fluctuations. As Q_l is relatively small and varies insignificantly, its effect on the input flow rate is minimal.

When $Q_{pv} > 0$:

$$Q_i + Q_{e1} + Q_{e2} > 0$$

$$Q_t = Q_i + Q_{e1} > 0$$

Even if, at this moment, Q_{pv} is small and Q_{e1}, Q_{e2} are negative (indicating an increase in pressure, which manifests as oil charging and storage flow), it still indicates that the inlet flow rate of the hydraulic impact mechanism is positive, flowing into the internal part of the impact mechanism. Conversely, when $Q_{pv} < 0$, meaning the piston acts as a pump during braking, we have:

$$Q_i + Q_{e1} + Q_{e2} < 0$$

$$Q_t = Q_i + Q_{e1} < 0$$

That is the phenomenon of "backflow," where the hydraulic impact mechanism expels oil from the inside to the outside, occurs—something generally considered unlikely. This negative flow phenomenon can be clearly observed in the measured waveform of the inlet flow rate in Fig. 7.2.

The experimental results have shown that compensation flow rate can also fully act as an 'accumulator.' But does the theoretical analysis yield the similar results?

Using the previously established mathematical and simulation models describing the motion process of the hydraulic impact mechanism without a high-pressure accumulator, a simulation program was developed, as listed in Appendix C. Figs. 7.3 and 7.4 are the results of the simulation calculations based on this program. Below are the output results of the simulation.

E_i	373.24	f	30.31				
η	0.47	η_m	0.698	η_v	0.69	η_p	0.84
P_{max}	15.11	P_{min}	8.86	P_{ave}	12.08		
u_{rm}	4.39	u_{im}	7.59				
T_r	18.120	T_b	3.68	T_i	10.52		
T_{vr}	2.08	T_{vi}	2.58				
S_p	47.29	S_r	39.16	S_b	8.13		
S_c	34	S_{ic}	7				
Q_i	120.85	P_s	6	P_{la}	0.7	k	0
M_p	12.95	M_v	0.3	V_{ko}	2572.63	V_{la}	450
A_1	14.78	A_2	5.02	A_3	1.53	A_5	3.13

By comparing Fig. 7.3 and its corresponding output results with Fig. 6.5 and the corresponding results, it can be seen that the impact frequency under both operating conditions has almost no change, and the impact energy is similar to the previous measured results. That is, when there is no high-pressure accumulator, the impact energy slightly decreases, and the efficiency drops more significantly. However,

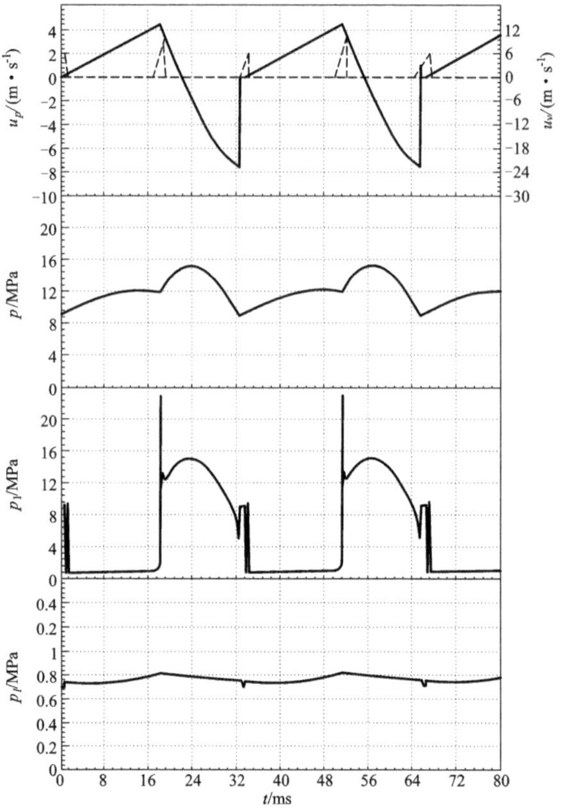

Fig. 7.3 Piston speed u_p, valve speed u_v (dashed line), inlet pressure p, rear chamber pressure p_1, and return pressure p_l without high-pressure accumulator

the inlet pressure fluctuations are much greater in the case without a high-pressure accumulator.

By comparing Fig. 7.4 with Fig. 6.6, it can be seen that although the flow rate variation curves required for piston and valve movement are almost the same, the variation in the inlet flow rate curves differs significantly.

The drastic variation in the flow rate Q_{pv} is mainly compensated by the flow rate Q_a generated from the volume change of the high-pressure accumulator when there is one. The variation in the flow rate Q_a from the high-pressure accumulator, as shown in Fig. 7.5, closely follows the changes in the flow rate Q_{pv} shown in Fig. 6.6 and plays a primary compensatory role. On the other hand, the compensation flow rate Q_e generated by the compression of the medium and the expansion of the rubber hose, although following a similar trend, provides much less compensation compared to Q_a. Since the compensation flow primarily comes from within the hydraulic impact mechanism, the inlet flow rate Q_t remains relatively smooth.

As shown in Fig. 7.6, since the compensating flow rate Q_a from the accumulator is zero, the compensation for the required flow rate Q_{pv} relies solely on the flow rate Q_e generated by the volume change of the medium and the rubber hose. Therefore,

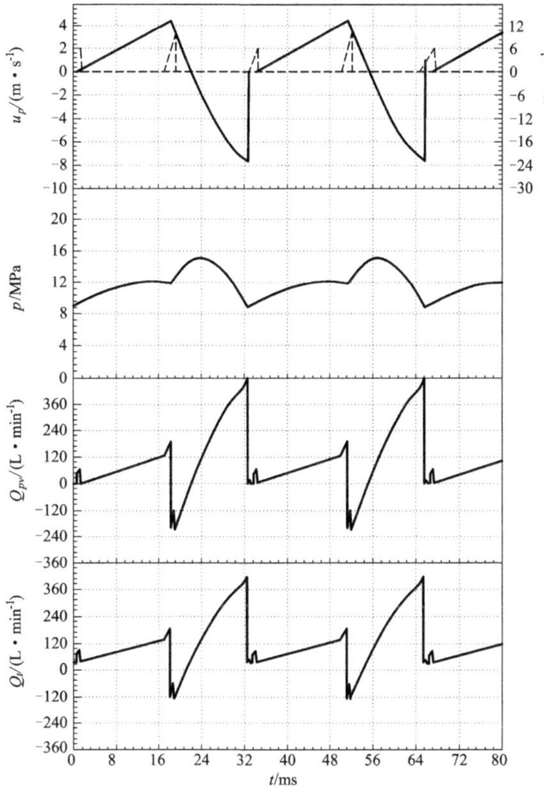

Fig. 7.4 Inlet pressure p, flow rate Q_{pv} required for piston and valve movement, and inlet flow rate Q_t without high-pressure accumulator

the variation of Q_e here is greater than that of Q_a in Fig. 7.5. Additionally, the compensating flow rate Q_{e2} generated by the internal medium volume V_{20} of the hydraulic impact mechanism is relatively small compared to Q_e. In other words, the compensating flow rate from the inside of the hydraulic impact mechanism to meet the requirements of Q_{pv} is relatively small, resulting in a more drastic variation in the inlet flow rate Q_t.

The above analysis, both practical and theoretical, proves that the compensation flow rate generated by the volume change of working medium and the transmission rubber hose in the hydraulic impact mechanism is significant and cannot be ignored. What are the main factors that influence this flow rate, and to what extent do they affect respectively? Below are the results of four sets of simulations for analyzing this issue. The simulation object is a hydraulic impact mechanism with an effective displacement q_p of nearly 50 mL (i.e., the effective flow rate consumed by the stroke piston movement). According to the grouping in Table 7.2, nine simulations were conducted. They are divided into four groups A, B, C, and D, based on the parameters V_{10}, V_{20}, K_l, and K_c related to the variation and compensation flow rate. Each group had three variations, and the second variation represents the intermediate value for comparison between groups. The values of V_{10}, V_{20}, K_l, and K_c are the same for all

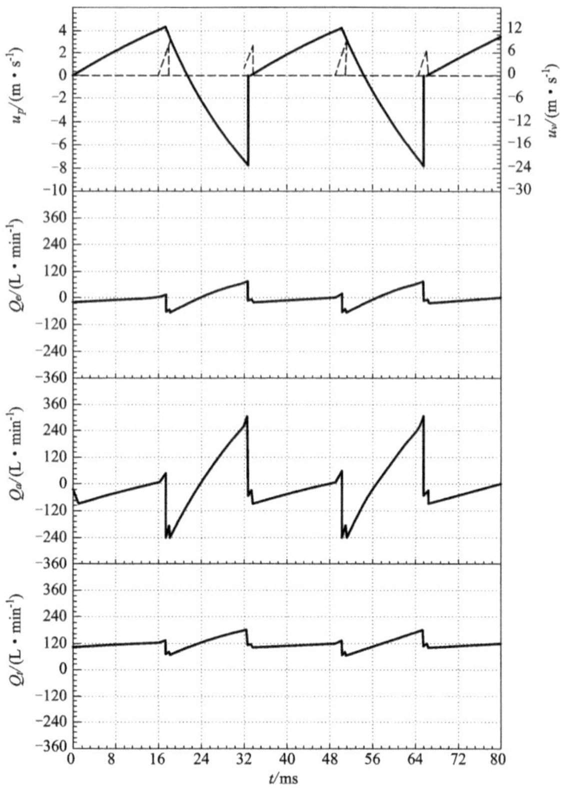

Fig. 7.5 Compensation flow rate Q_e, accumulator flow rate Q_a and inlet flow rate Q_t with a high-pressure accumulator

groups. The simulations were conducted under the condition without a high-pressure accumulator, calculating with the program from Appendix C.

Besides depicting the impact of V_{10}, V_{20}, K_l, and K_c on inlet pressure p, compensation flow rates Q_e and Q_{e2}, and inlet flow rate Q_t in Figs. 7.7, 7.8, 7.9 and 7.10, the results of the nine simulations are summarized numerically in Table 7.3 to reflect the influence of these parameters on several key performance and operational parameters of the hydraulic impact mechanism.

Upon analyzing and observing Tables 7.2 and 7.3, as well as Figs. 7.7, 7.8, 7.9 and 7.10, the following observations can be made:

(1) An increase in V_{10} and V_{20}, as well as decrease in K_l and K_c, all lead to a smoother inlet pressure p.
(2) A decrease in V_{10} and an increase in V_{20} will result in a smoother variation in the inlet flow rate Q_t, while the compensation flow rate Q_{e2} will experience a larger variation amplitude.
(3) V_{10}, V_{20}, K_l and K_c have little effect on the total compensation flow rate Q_e, and the latter two have minimal effect on the variation of the inlet flow rate Q_t.
(4) The parameters with the greatest impact on pressure and flow rate are the volume of the high-pressure rubber hose V_{10} and its bulk modulus of elasticity K_c.

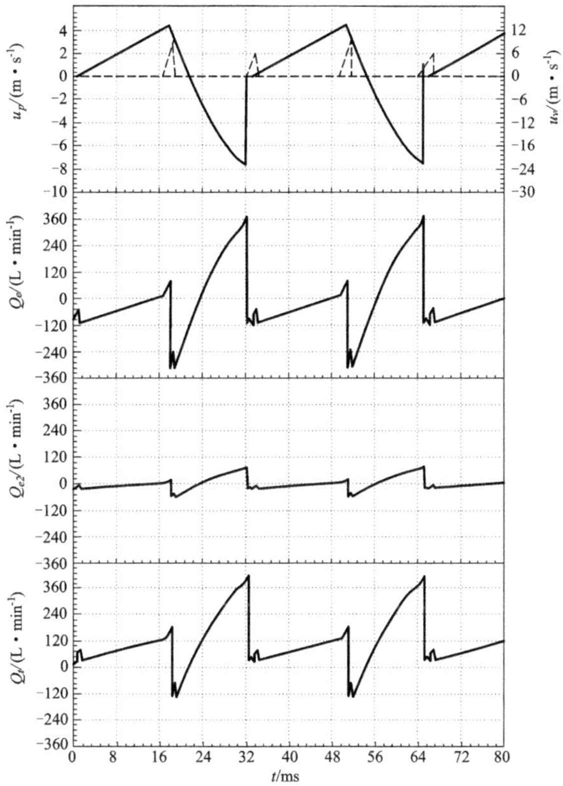

Fig. 7.6 Compensation flow rate Q_e, compensation flow rate Q_{e2} generated by the compression of the medium inside the impact mechanism, and inlet flow rate Q_t without a high-pressure accumulator

Table 7.2 Parameter values for the corresponding curves in Figs. 7.7, 7.8, 7.9 and 7.10

	A Change V_{10} (Fig. 7.7}	B Change V_{20} (Fig. 7.8)	C Change K_l (Fig. 7.9)	D Change K_c (Fig. 7.10)
1	V_{10}= 500 mL The rest are the same as in Group B (K_e = 397 MPa)	V_{20} = 500 mL The rest are the same as in B (K_e = 152 Mpa)	K_l = 1500 MPa The rest are the same as in B (K_e = 270 MPa)	K_c = 450 MPa The rest are the same as in B (K_e = 330 MPa)
2	V_{10} = 1000 mL, V_{20} = 1000 mL, K_l = 1000 MPa, K_c = 300 MPa (K_e = 248 MPa)			
3	V_{10} = 2000 mL The rest are the same as in Group B (K_e = 141 MPa)	V_{20} = 2000 mL The rest are the same as in B (K_e = 386 MPa)	K_l = 700 MPa The rest are the same as in B (K_e = 224 MPa)	K_c = 200 MPa The rest are the same as in B (K_e = 180 MPa)

Note V_{10}: high-pressure hose volume corresponding to 10 q_p, 20 q_p, 40 q_p, respectively;
V_{20}: volume of the high-pressure chamber inside the impact mechanism, corresponding to 10 q_p, 20 q_p, and 40 q_p, respectively;
K_l: Bulk modulus of elasticity of the working medium;
K_c: Bulk modulus of elasticity of the high-pressure hose

Fig. 7.7 Effect of high-pressure hose volume V_{10} on inlet pressure p, compensation flow Q_e, and Q_{e2}, and inlet flow Q_t

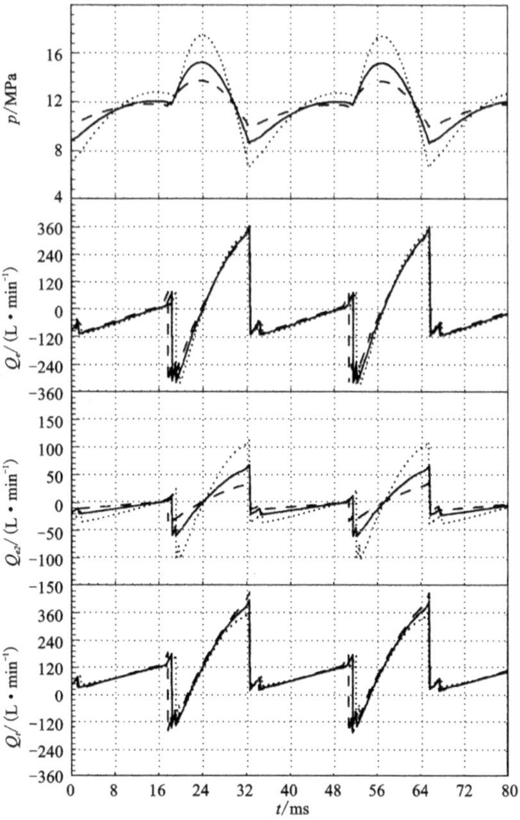

(5) These four parameters have minimal influence on the impact energy E_i, impact frequency f, and the time-averaged pressure p_{ave}.

Based on the above analysis, the following conclusions can be drawn:

(1) The variations in the volume of the high-pressure rubber hose and the compression rate of the working medium in the high-pressure chamber of the hydraulic impact mechanism are essential factors to consider in research and design.
(2) The high-pressure rubber hose and the working medium in the high-pressure chamber essentially act as an implicit high-pressure accumulator with high stiffness. The compensation flow provided by them can maintain the normal operation of the hydraulic impact mechanism even without a high-pressure accumulator. When the high-pressure accumulator is absent, the impact frequency remains nearly unchanged, and the decrease in impact energy and efficiency is not significant.
(3) The high-pressure inlet flow of the hydraulic impact mechanism is pulsating, even when the pump flow rate is constant.

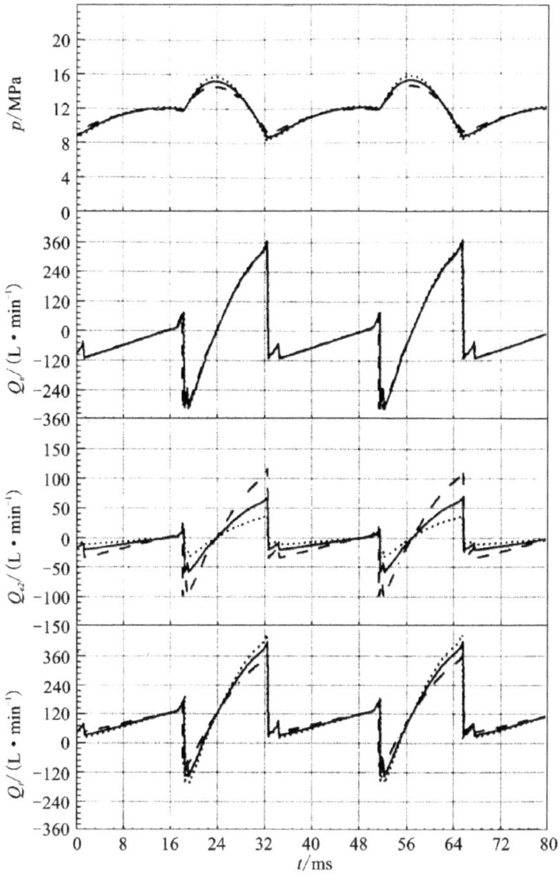

Fig. 7.8 Effect of high-pressure oil chamber volume V_{20} inside the impact mechanism on inlet pressure p, compensation flow Q_e, and Q_{e2}, and inlet flow Q_t

(4) The volume of the high-pressure rubber hose and its bulk modulus of elasticity have the most significant influence on the fluctuation amplitude of the inlet pressure p and inlet flow rate Q_t.

It should be noted that, for the sake of simplicity, the author did not consider the length effect of the high-pressure rubber hose when analyzing its compensating effect due to volume change. The analysis did not take into account the distribution parameters, neglecting the phase lag caused by fluid capacitance and fluid compliance during pressure wave transmission, and the response between compensation flow rate variation and the piston's movement frequency in the hydraulic impact mechanism. Therefore, in the simulation process, the entire length of the high-pressure rubber hose was not included; only a shorter section, generally not exceeding 3 m, was considered. Although this approach led to some distortion between the simulated inlet flow rate variation curve and the measured waveform, the overall variation trend remains consistent.

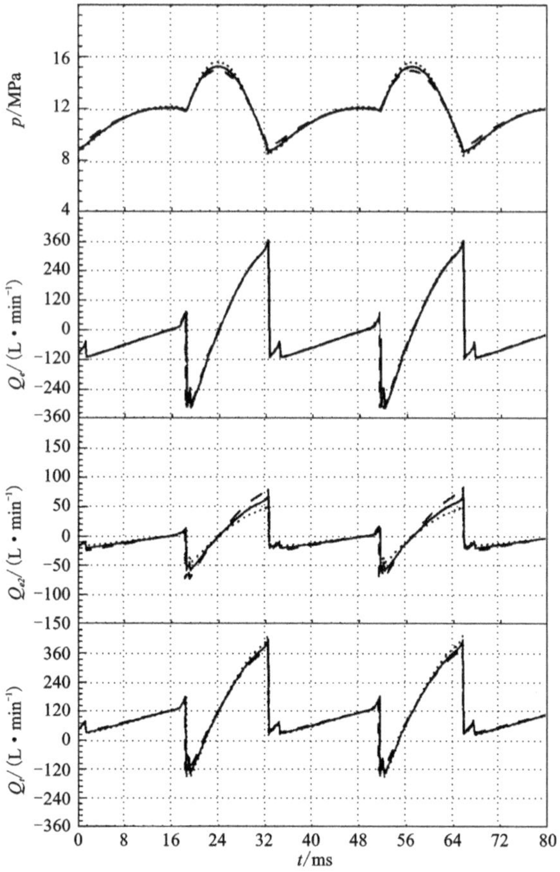

Fig. 7.9 Effect of the Bulk Modulus of Elasticity of the Working Medium K_l on inlet pressure p, compensation flow Q_e, and Q_{e2}, and inlet flow Q_t

7.2 Return Oil and Return Oil Accumulator in Hydraulic Impact Mechanisms

In the previous section, a detailed study was conducted on the high-pressure inlet flow rate of the hydraulic impact mechanism. In this section, we will discuss the essence of the low-pressure outlet flow rate, which is the return oil flow rate, and the role of the return oil accumulator. Currently, there is a lack of comprehensive research on return oil accumulators in the industry, and there has been no in-depth discussion of their working principles or scientifically established parameter design methods. Here, based on an analysis of return oil pressure and the causes of return oil cavitation in the hydraulic impact mechanism, a parameter design method for the return oil accumulator is proposed, establishing its essential role in the hydraulic impact mechanism.

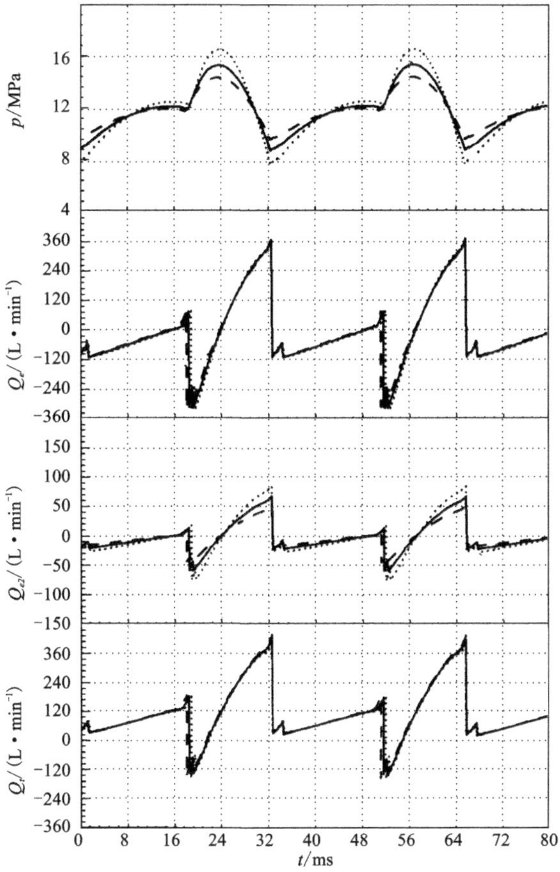

Fig. 7.10 Effect of bulk modulus of elasticity of high-pressure hose K_c on inlet pressure p, compensation flow Q_e, and Q_{e2}, and inlet flow Q_t

Table 7.3 Effect of V_{10}, V_{20}, K_l, and K_c on impact energy, frequency, and pressure change

	E_i /J	f /Hz	P_{max} /MPa	P_{min} /MPa	P_{ave} /MPa
A2, B2, C2, D2	371.87	30.21	15.25	8.65	12.03
A1	376.65	30.28	17.44	6.66	12.31
A3	369.77	30.23	13.77	10.04	11.93
B1	373.93	30.27	15.73	8.26	12.12
B3	370.1	30.18	14.59	9.23	11.96
C1	373.41	30.26	15.59	8.38	12.1
C3	372.04	30.26	14.93	8.98	12.03
D1	374.72	30.23	16.47	7.56	12.18
D3	370.18	30.22	14.29	9.53	11.95

1. **Return Oil Resistance and Cavitation**

In a typical hydraulic system, the causes of return oil resistance, aside from intentionally set backpressure, generally include pipeline hydraulic resistance, throttle losses in filters, and local losses. However, if we consider the unique characteristics of hydraulic impact mechanisms compared to standard hydraulic machinery, we will find that there are additional, more significant causes of return resistance oil.

As pointed out in Chap. 1 of this book, the working oil pressure in hydraulic impact mechanisms mainly depends on the inertia forces of their own movements, which distinguishes them from typical hydraulic machinery where the working oil pressure primarily depends on external loads. The piston, valve spool, and other moving components continuously undergo variable-speed motions with accelerations as high as 400–2000 m/s^2, and this is the reason for this particular characteristic. Consequently, it can be inferred that the oil, as it experiences such intense variable speed movements along with the piston and valve spool, will naturally generate a certain inertia resistance. This inertia force is one of the volume forces in unsteady fluid flow. Clearly, when the mass of the oil relative to the piston and valve spool reaches a considerable value, the oil pressure resulting from inertia force can also attain significant values.

The return oil pipes of hydraulic impact mechanisms are typically relatively long, which means that the mass of the hydraulic oil is also relatively large. When the control chamber of the piston or valve discharges oil into the return oil pipe, the oil accelerated by their movement generates significant inertial oil pressure, which hinders the motion of the piston or valve spool.

Taking the YYG250 hydraulic rock drill impact mechanism with a constant-pressure front chamber and external control valve, designed by Central South University, (shown in Fig. 4.1), as an example, we will analyze and calculate its inertial oil return resistance.

Figure 7.11 illustrates the simultaneous discharge of oil into the return pipe by the piston and valve. The return oil pressure p_l is primarily governed by the combined motion of the piston (with mass is m_p), valve spool (m_v), and oil in return pipe (m_0). Since the volume of oil discharged by the piston and valve in one cycle is much smaller than the volume of oil in the return pipe, m_0 can be considered constant during the discharge process. The oil pressure resulting from the backpressure of the oil supply, filters, and coolers, concentrated at the end of the return pipe, is denoted as p_f, while other resistances are neglected. In the absence of a return oil accumulator, the following equations can be derived.

Based on the forces acting on the piston and valve spool, the following can be established:

$$m_p \frac{d^2 y_p}{dt^2} = A_2 p - A_1 p_l \tag{7.4}$$

$$m_v \frac{d^2 y_v}{dt^2} = A_3 (p - p_l) \tag{7.5}$$

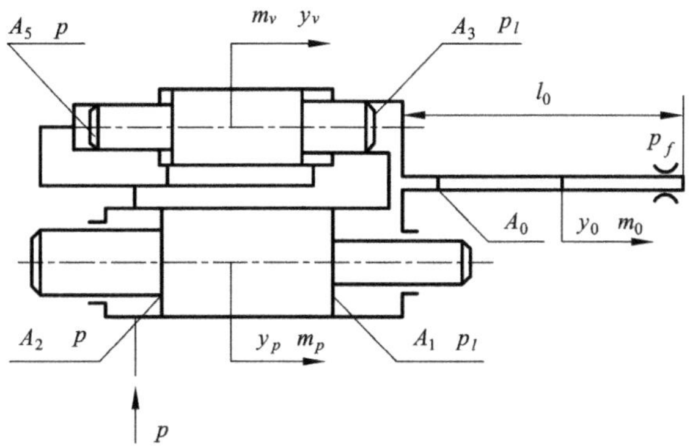

Fig. 7.11 Schematic diagram of simultaneous oil discharge by the piston and valve
m_p, m_v, m_0: mass of the piston, valve spool, oil in return line, respectively;
y_p, y_v, y_0: displacement of m_p, m_v, and m_0, respectively;
A_1, A_2, A_3, A_4: axial oil pressure acting areas of piston and valve spool;
A_0, l_0: cross-sectional area and length of the return line;
S_v: Stroke of the valve; p: Supply oil pressure;
Q: Oil supply flow rate; p_l: return pressure

$$m_0 \frac{d^2 y_0}{dt^2} = A_0 (p_l - p_f) \tag{7.6}$$

Neglecting the compressibility of the working medium, according to the principle of continuity for fluids, we have:

$$A_1 y_p + A_3 y_v = A_0 y_0 \tag{7.7}$$

Taking the second derivative with respect to time on both sides of Eq. (7.7), we get:

$$A_1 \frac{d^2 y_p}{dt^2} + A_3 \frac{d^2 y_v}{dt^2} = A_0 \frac{d^2 y_0}{dt^2} \tag{7.8}$$

With the substitution of the expressions for the acceleration of the piston, valve, and return oil, $\frac{d^2 y_p}{dt^2}$, $\frac{d^2 y_v}{dt^2}$, and $\frac{d^2 y_o}{dt^2}$, derived from Eqs. (7.4)–(7.6), into Eq. (7.8), the solution can be derived:

$$p_l = \frac{m_0 \left(m_v A_1 A_2 + m_p A_3^2 \right) p + m_p m_v A_0^2 p_f}{m_p m_v A_0^2 + m_v m_0 A_1^2 + m_0 m_p A_3^2} \tag{7.9}$$

Similarly, the solution can be derived for the case when only the valve discharges oil:

$$p_l = \frac{m_0 A_3^2 p + m_v A_0^2 p_f}{A_0^2 m_v + A_3^2 m_0} \tag{7.10}$$

When only the piston discharges oil:

$$p_l = \frac{A_1 A_2 m_0 p + A_0^2 m_p p_f}{A_0^2 m_p + A_1^2 m_0} \tag{7.11}$$

Figure 7.12a shows the velocity curve of the piston u_p, with u_{rm} and u_{im} denoting the maximum velocities during the return stroke and impact strokes of the piston, respectively. Figure 7.12b represents the variation curve of the oil flow rate Q_{io} entering the return pipe (assuming that the valve completes its impact stroke reversal just before the start of the next return stroke). Let Q_p, Q_v, and Q_l represent the discharge flow rates from the piston, the control chamber of the valve, and leakage, respectively. Each segment of the curve in Fig. 7.12b represents the following flow rates:

$$Q_{ab} = Q_p + Q_l \qquad Q_{cd} = Q_l \qquad Q_{de} = Q_l + Q_v$$

The variation of flow rate Q_0 in the return pipe is represented in Fig. 7.12c. During one impact cycle, although the flow rate curves in Fig. 7.12b, c have equal integrals over time, the changes occur in different rates. When the flow rate entering the return pipe cannot meet the flow rate demand in the pipe, cavitation occurs. Let the cavitation generated during one impact cycle (during the piston braking and impact time t_c as

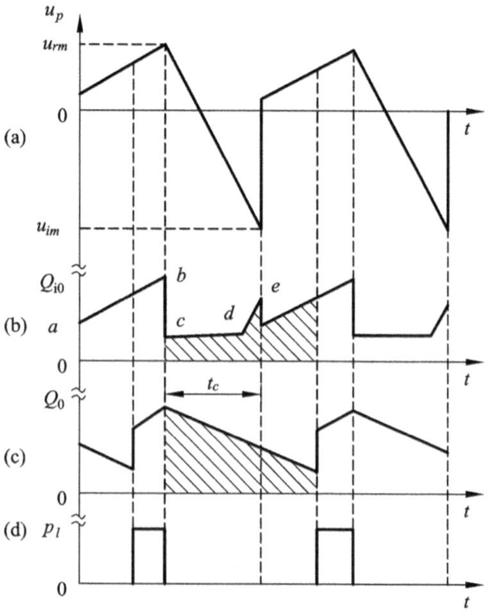

Fig. 7.12 Return oil flow and pressure variation without a return oil accumulator

shown) be denoted as V_c. Assuming the oil in the return pipe undergoes uniform deceleration under the effect of oil pressure p_f during this period, the cavitation volume is given by:

$$V_c = A_0\left(u_{cr}t_c - \frac{1}{2}a_0t_c^2\right) - Q_lt_c - A_3S_v \tag{7.12}$$

In the equation:

the magnitude of the negative acceleration of the return oil is $a_0 = \frac{A_0p_f}{m_0}$;

the velocity of the return oil at the end of the piston return stroke acceleration is $u_{cr} = \frac{A_1u_{rm}+Q_l}{A_0}$;

S_v is the stroke of the valve.

Equation (7.12) then becomes:

$$V_c = A_1u_{rm}t_c - \frac{A_0^2p_f}{2m_0}t_c^2 - A_3S_v \tag{7.13}$$

Only when the cavitation in the return oil is refilled can the inertial return oil pressure calculated by Eqs. (7.9)–(7.11) be generated. In general, the cavitation in the return oil can only be refilled during the later phase of the piston's return stroke acceleration, resulting in the generation of pulsating inertial return oil pressure [see Fig. 7.12d]. At this time, the shaded areas in Fig. 7.12b, c are equal.

With the increasing the sensitivity of the testing instruments, it was found that the return oil pressure p_l is almost zero for most of the time (see Fig. 7.13), indirectly proving the existence of oil cavitation in the return flow and explaining why the return oil pressure appears as pulses. Clearly, the pulse width of the inertial pressure depends on the size of the cavitation. In Eq. (7.13), the first and third terms on the right side, as well as the values of A_0, t_c, and m_0, are constant for a given impact mechanism. Therefore, as the oil pressure p_f increases, the cavitation volume V_c decreases, and the pulse width increases. This was confirmed in experiments by artificially adjusting the oil pressure p_f, which induced changes in the inertial oil pressure.

Figure 7.13 shows the measured waveforms of the front chamber oil pressure p and the return oil pressure p_l. In fact, the intensity of the pulsating inertial oil pressure is calculation by Eq. (7.11). Considering that $p_f \ll p$, Eq. (7.11) becomes:

$$p_l = \frac{A_1A_2m_0}{A_0^2m_p + A_1^2m_0}p = \frac{A_1A_2}{A_1^2 + \frac{A_0m_p}{\rho l_0}}p. \tag{7.14}$$

In the equation:

ρ: density of the fluid;

l_o: length of the return oil pipe.

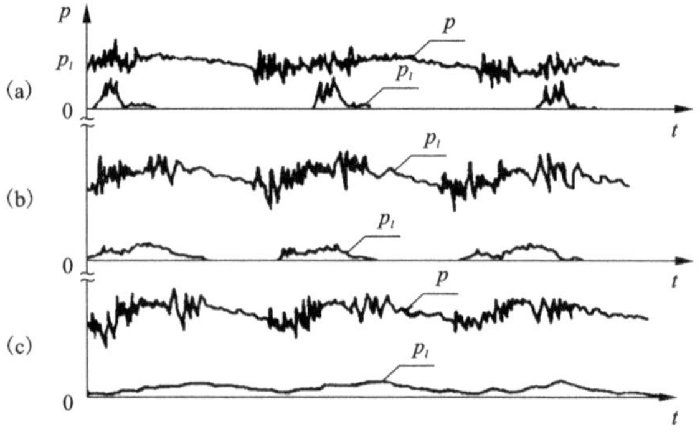

Fig. 7.13 Measured waveforms of p and p_l

The ratio of the resistance formed by the inertial return oil pressure to the front chamber oil pressure:

$$\eta = \frac{p_l A_1}{p A_2} = \frac{1}{1 + \frac{A_0 m_p}{\rho l_o A_1^2}} \tag{7.15}$$

Equation (7.15) indicates that the longer the return oil pipe, especially with a larger rear chamber area of the piston, the greater the return oil resistance. Conversely, increasing the flow area of the return oil pipe and increasing the piston's mass will reduce the return oil resistance.

Figure 7.13 shows the waveforms of the front chamber oil pressure and return oil pressure obtained during tests with two pistons of similar mass but different areas A_1 and A_2. With the calculation of actual parameters in the Eq. (7.14), $p_l = 0.431p$ in Fig. 7.13a and $p_l = 0.1315p$ in Fig. 7.13b. The measured return oil pressure p_l and the front chamber oil pressure p closely follow this relationship, confirming the accuracy of the above analysis.

2. **Return Oil Accumulator**

The inertial return oil pressure and return oil cavitation are undoubtedly harmful to hydraulic impact mechanisms. They disrupt the continuity of the oil flow, generate intense vibrations, increase power consumption, and lead to early damage of the oil pipes. On-site observations have shown that some hydraulic rock drills lacking a return oil accumulator, experience more severe damage to their return oil pipes than to their high-pressure inlet pipes.

From the analytical formula (7.15), the coefficient η can be reduced by changing parameters. Among several options such as reducing the rear chamber area A_1, increasing the cross-sectional area A_0 of the return oil pipe, shortening the length l_o of the return oil pipe, and increasing the piston's mass m_p, reducing the rear chamber area A_1 is the most feasible and effective method. However, reducing A_1 is closely related to the overall design.

The return oil pressure in Fig. 7.13b is much lower than that in Fig. 7.13a, which is the result of changing parameters such as the area A_1. However, the difference in coefficient η is not significant; in Fig. 7.13a, $\eta = 0.691$, and in Fig. 7.13b, $\eta = 0.435$. Additionally, the selection of A_1 is constrained by other factors and cannot solve the issue of return oil cavitation. The following analysis indicates that setting a reasonably designed return oil accumulator can simultaneously reduce the inertial return oil pressure and eliminate return oil cavitation.

Neglecting leakage, other resistance, and losses, the return oil pressure p_l during piston oil discharge after setting a return oil accumulator is determined by the following equations:

$$m_p \frac{d^2 y_p}{dt^2} = A_2 p - A_1 p_l, \tag{7.16}$$

$$m_0 \frac{d^2 y_0}{dt^2} = A_0 (p_l - p_f), \tag{7.17}$$

$$p_l V_{ll}^k = C_l = p_{al} V_{al}^k, \tag{7.18}$$

$$A_1 y_p = A_0 y_0 + V_i - V_{ll}. \tag{7.19}$$

where k: gas variability index;

p_{al}, V_{al}, V_i, and V_{ll}: represent the inflation pressure, inflation volume, instantaneous working volume, and initial working volume of the return oil accumulator, respectively.

By simultaneously solving Eqs. (7.16)–(7.19), a differential equation for the return oil pressure p_l during piston oil discharge can be obtained:

$$\begin{aligned} \frac{d^2 y_l}{dt^2} &- \frac{1+k}{k} p_l^{-1} \left(\frac{dp_l}{dt} \right)^2 + k C_l^{-\frac{1}{k}} \left(\frac{A_0^2}{m_0} + \frac{A_1^2}{m_p} \right) p_l^{\frac{1+2k}{k}} \\ &- k C_l^{-\frac{1}{k}} \left(\frac{A_0^2}{m_0} p_f + \frac{A_1 A_2}{m_p} \right) p_l^{\frac{1+k}{k}} = 0. \end{aligned} \tag{7.20}$$

When the piston does not discharge oil, the return oil pressure is mainly determined by the equation:

$$m_0 \frac{d^2 y_0}{dt^2} = A_0 (p_l - p_f), \tag{7.21}$$

$$p_l V_{ll}^k = C_l, \tag{7.22}$$

$$-A_0 y_0 = V_i - V_{ll}. \tag{7.23}$$

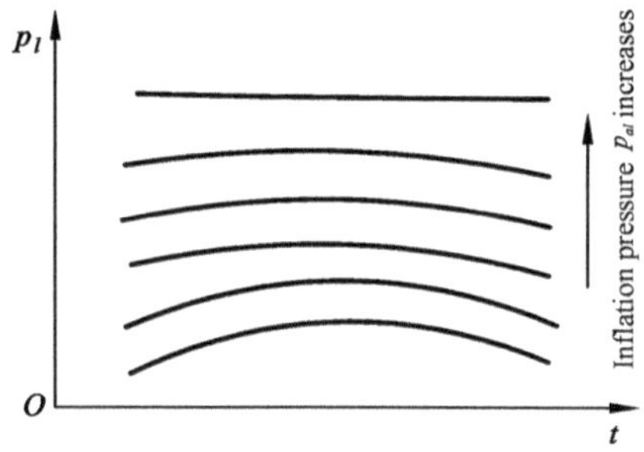

Fig. 7.14 Relationship between p_l and p_{al}

They also can be changed to a differential equation for the return pressure p_l when the piston doesn't discharge oil.

$$\frac{d^2y_l}{dt^2} - \frac{1+k}{k}p_l^{-1}\left(\frac{dp_l}{dt}\right)^2 - \frac{kA_0^2}{m_0}C_l^{-\frac{1}{k}}p_l^{\frac{1+2k}{k}} + \frac{kA_0}{m_0}C_l^{-\frac{1}{k}}p_f p_l^{\frac{1+k}{k}} = 0. \quad (7.24)$$

Equations (7.20) and (7.24), two nonlinear differential equations, respectively describe the variation of return oil pressure (accumulator gas pressure) during oil charging and discharging of the return oil accumulator. Treating p and p_f as constants, these two nonlinear differential equations can be solved numerically with a computer. The solutions reveal that:

The value of $C_l = p_{al}V_{al}^k$ has a significant effect on the return oil pressure p_l. As the inflation pressure p_{al} increases, the return oil pressure p_l also rises accordingly (see Fig. 7.14). When the value of p_{al} exceeds the return oil pressure determined by Eq. (7.14), the return oil accumulator ceases to function. Reducing V_{al} or increasing p_{al} yields similar effects, but the latter has a greater impact than the former. Experimental results also confirm these conclusions.

How to reasonably determine the parameters for the return oil accumulator? Hydraulic impact mechanisms in rock drilling machines typically use diaphragm-type accumulators, as shown in Fig. 7.15, where the upper and lower volumes are roughly equal. During normal operation, the diaphragm's deviation from the central position (III) are essentially the same. In the figure, I, II, and IV represent the extreme positions of the diaphragm during inflation, oil discharge, and oil charging, respectively. p_{al}, V_{al}, p_{ll}, V_{ll}, p_{lh}, and V_{lh} indicate the corresponding states' gas pressure and gas volume. ΔV_a represents the accumulator's displacement volume.

For a return oil accumulator with reasonable parameters, the variation of the oil velocity u_0 in the return oil pipe over one cycle is shown in Fig. 7.16a–c show the corresponding variation curves for the accumulator's flow rate (dV_i/dt) and pressure p_l (with flow rate discharged from the accumulator considered positive). When $p_l > p_f$, the return oil accelerates, and when $p_l < p_f$, the return oil decelerates. We define the condition where $p_l = p_f$ as the diaphragm being in the middle position III. In

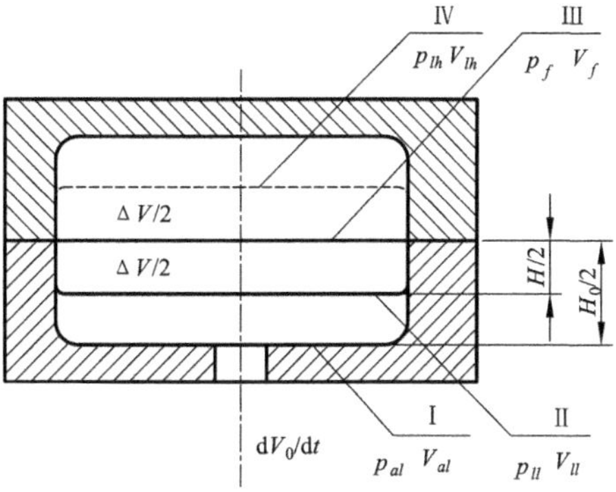

Fig. 7.15 Design parameters of the return oil accumulator

Fig. 7.16b, the moments labeled as *b* and *d* represent the diaphragm at the middle position, while the moments *a* and *c* represent the diaphragm at positions II and IV, respectively.

In the gas state equation of the accumulator,

$$p_{ll} V_{ll}^k = p_{lh} V_{lh}^k, \tag{7.25}$$

the ratio of the maximum working pressure to the minimum working pressure is defined as $\frac{p_{lh}}{p_{ll}} = \lambda$, then we have:

$$p_{ll}\left(\frac{V_{al}}{2} + \frac{\Delta V_a}{2}\right)^k = \lambda p_{ll}\left(\frac{V_{al}}{2} - \frac{\Delta V_a}{2}\right)^k. \tag{7.26}$$

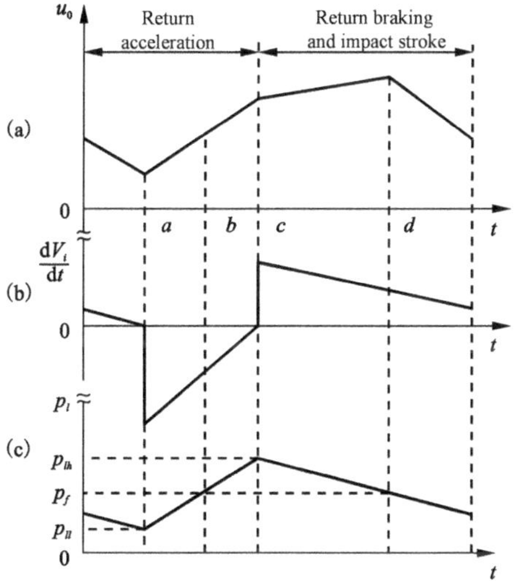

Fig. 7.16 Flow rate and pressure variation diagrams of the return oil with an accumulator

The inflation volume is obtained as:

$$V_{al} = \frac{\lambda^{\frac{1}{k}} + 1}{\lambda^{\frac{1}{k}} - 1} \Delta V_a. \tag{7.27}$$

After selecting the cross-sectional area S of the accumulator (which is generally chosen to be as large as possible within the allowable installation space to minimize diaphragm deformation), the height H_a of the accumulator in Fig. 7.15 can be more precisely determined,

$$H_a = \frac{\lambda^{\frac{1}{k}} + 1}{\lambda^{\frac{1}{k}} - 1} H, \tag{7.28}$$

where, $H = \frac{\Delta V_a}{S}$.

By using the equations

$$p_f V_f^k = p_{al} V_{al}^k, \tag{7.29}$$

which can be written as

$$p_f \left(\frac{V_{al}}{2}\right)^k = p_{al} V_{al}^k. \tag{7.30}$$

The inflation pressure can also be obtained as:

$$p_{al} = 2^{-k} p_f. \tag{7.31}$$

The backpressure p_f is generated by components such as the filter, cooler, and the replenishment pressure of the closed-loop system. Its value typically ranges from 0.5 to 1.5 MPa.

Substituting the piston's return acceleration stroke S_r and the corresponding displacement y_{0r} of the center of mass of the return oil into Eq. (7.19), the accumulator's charging volume (i.e., displacement) can be obtained:

$$\Delta V_a = V_{lh} - V_{ll} = A_1 S_r - A_0 y_{0r}. \tag{7.32}$$

The value of y_{0r} in Eq. (7.32) is difficult to calculate and determine. A differential equation for the displacement y_0 of the return oil's center of mass can be derived from Eqs. (7.16)–(7.19).

$$\frac{d^4 y_0}{dt^4} + \frac{(1+k)m_0}{kA_0}\left(\frac{m_0}{A_0} \cdot \frac{d^2 y_0}{dt^2} + p_f\right)^{-1}\left(\frac{d^3 y_0}{dt^3}\right)^2 - \frac{kA_0^2}{m_0} C_l^{-\frac{1}{k}}\left(\frac{m_0}{A_0} \cdot \frac{d^2 y_0}{dt^2} + p_f\right)^{\frac{1+k}{k}} \frac{d^2 y_0}{dt^2}$$

$$+\frac{kA_0A_1}{m_0m_p}C_l^{-\frac{1}{k}}\left(A_2p - A_1p_f - \frac{A_1m_0}{A_0}\cdot\frac{d^2y_0}{dt^2}\right)\left(\frac{m_0}{A_0}\cdot\frac{d^2y_0}{dt^2} + p_f\right)^{\frac{1+k}{k}} = 0 \quad (7.33)$$

Treating p and p_f as constants, within the time of the piston's return acceleration, solving this high-order nonlinear equation numerically will yield y_{0r}. Substituting the value of y_{0r} into Eq. (7.32) will give ΔV_a. However, this is a rather complicated task. In typical engineering design, a more convenient method must be sought.

From Fig. 7.16, it can be observed that the displacement of m_0 during the piston's return acceleration phase is certainly smaller than the displacement y_{0i} during the piston's return braking and impact phases. Adopting a conservative design approach, let $y_{0r} = y_{0i}$. According to Eq. (7.23), it can be inferred that

$$\Delta V_a = V_{ll} - V_i = A_0y_{0i} = A_0y_{0r} \quad (7.34)$$

By simultaneously solving Eqs. (7.32) and (7.34), we obtain

$$\Delta V_a = \frac{1}{2}A_1S_r \quad (7.35)$$

The value of λ directly affects the inflation volume V_{la} and the fluctuation degree of the return oil pressure. Generally, λ = 1.2–1.5 (with K = 1.4). This leads to a more specific calculation formula,

$$V_{al} = (15 \sim 7)\Delta V_a = (7.5 \sim 3.5)A_1S_r \quad (7.36)$$

$$p_{al} = 0.36p_f \quad (7.37)$$

Obviously, compared to high-pressure accumulators, the return oil accumulator designed using the above formula can significantly reduce both its volume and shell thickness. It is worth mentioning that the same method can also be used to design the inflation volume and inflation pressure of high-pressure accumulators, as shown in Fig. 7.15, where the diaphragm divides the inflation volume into equal halves.

The author used the aforementioned formula to design a return oil accumulator with $V_{al} = 3.3 \times 10^{-5}$ m^3 and $p_{al} = 0.3$ Mpa, replacing the original accumulator with $V_{al} = 8.7 \times 10^{-5}$ m^3 in the YYG90 impact mechanism, achieving good results. Fig. 7.13c displays the measured waveform of the oil pressure in the front chamber of the piston and the return oil pressure (accumulator gas pressure). The return oil pressure waveform changes smoothly, and the continuous operation of the accumulator indicates that return oil cavitation has been eliminated. The vibration of the return oil pipe has significantly reduced (for clarity, the amplitude of p_l in Fig. 7.13c has been magnified by 4 times relative to p).

The above discussion theoretically explains the inertial return oil pressure and return oil cavitation in the constant front chamber pressure type hydraulic impact mechanism. It highlights that the essential function of the return oil accumulator is to reduce the inertial return oil pressure and eliminate return oil cavitation. Additionally, a concise and scientifically sound formula for the parameter design of the return oil accumulator has been proposed.

The above discussion theoretically explains the inertial return oil pressure and return oil cavitation in the constant front chamber pressure hydraulic impact mechanism. It points out that the essential function of the return oil accumulator is to reduce inertial return oil pressure and eliminate return oil cavitation. Additionally, a concise and scientific design formula for the parameters of the return oil accumulator is proposed.

Furthermore, it's worth noting that the variation process of the return oil flow rate and return oil pressure without the installation of a return oil accumulator are highly complex. To obtain clearer analytical results, a simplified mathematical model has been used to describe these variations.

7.3 Control Valve and Pressure Surges and Cavitation of Hydraulic Impact Mechanisms

The control valve of a hydraulic impact mechanism is a critical component, and there is substantial discussion regarding its research and design. The research in this study is based on the test curve of the rear chamber pressure p_1 of the hydraulic impact mechanism, as shown in Fig. 7.17.

From Fig. 7.17, it is evident that the rear chamber pressure p_1 of the hydraulic impact mechanism exhibits a high-value and short-duration pressure pulse when the pressure suddenly increases or decreases. The following research indicates that the generation of pressure pulses is related to the neutral position of the control valve or the timing of impact stroke reversal.

It is well known that hydraulic shocks and cavitation phenomena in hydraulic machinery are two harmful transient processes. During the movement of the control valve near the neutral position in a hydraulic impact mechanism, cavitation may occur in the control oil chamber, as well as oil pressure surges similar to the effects of hydraulic shocks. In this section, the causes of these phenomena and methods for their prevention will be explored.

Figures 7.18 and 7.19 respectively illustrate the working principles of two typical externally mounted slide-valve hydraulic rock drill impact mechanisms: the rear-control type (with a constantly high-pressure front chamber) and the dual-control type (with alternating pressure oil flow to both the front and rear chambers). The major difference between the two in terms of the control valve is that the former has only two control ports, z_1 and z_2, while the latter has four control ports, z_1, z_2, z_3, and z_4. In the figures, the valve spool is in the neutral position. The following analysis shows that a neutral position with zero opening or negative opening is completely unsuitable for hydraulic impact mechanisms, so only positive opening cases are studied here.

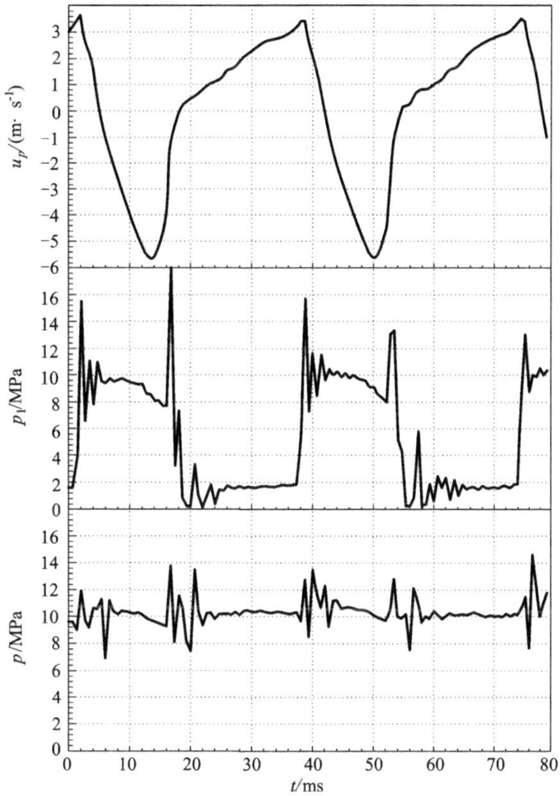

Fig. 7.17 Measured curves of piston speed u_p, rear chamber pressure p_1, inlet pressure p

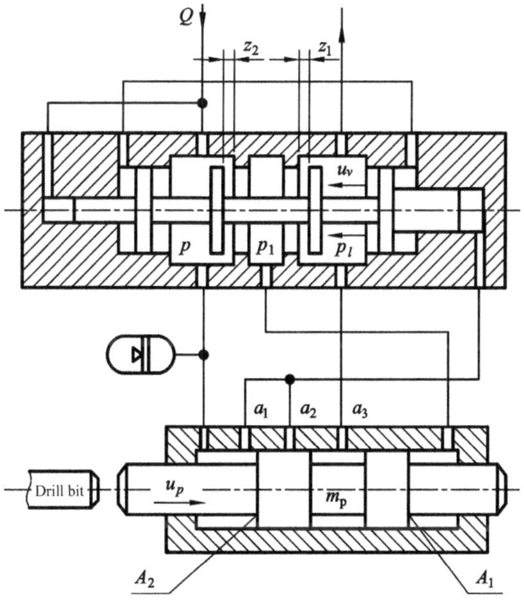

Fig. 7.18 Control ports of the control valve of the rear-control hydraulic impact mechanism

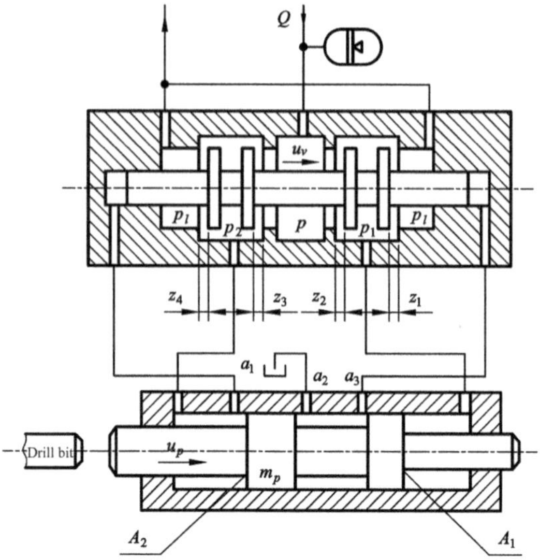

Fig. 7.19 Control port of the control valve of the dual-control hydraulic impact mechanism

In the figures, each valve opening represents an exaggerated positive opening. As the valve spool moves, the sizes of these valve ports change, causing the oil pressure in the piston chambers to vary due to throttling effects. Clearly, when the valve opening is too small while the flow rate of incoming (or outgoing) oil is relatively large, a significant oil pressure surge or cavitation may occur in the corresponding oil chamber. The analysis will proceed based on the stages of piston movement.

1. **Oil pressure surge at the end of the return acceleration phase**

Let's take the rear-control type hydraulic impact mechanism shown in Fig. 7.18 as an example. When the piston accelerates during the return stroke (moving to the right) and opens the valve's control oil port a_1, the valve spool reverses (moving to the left), causing the valve port z_1, which controls the return oil from the rear chamber of the piston, to gradually close. As the spool moves near the neutral position, the piston has reached its maximum return stroke velocity u_{rm}. If the valve has a zero opening or negative opening design, a brief moment of flow interruption will occur in the rear chamber, where a small portion of the piston's kinetic energy and the oil's kinetic energy in the chamber will be converted into oil pressure energy, causing a sharp increase in rear chamber oil pressure, resulting in hydraulic shock.

Generally, a positive opening valve is used to eliminate this harmful phenomenon. However, as the valve spool crosses the neutral position, both the spool and piston are moving at relatively high speeds. If the positive opening is too small, the oil 'squeezing' through the valve port will also generate a significant pressure surge in a very short time (within tens of microseconds). Figure 7.17 shows the rear chamber pressure as a square wave, with a high-amplitude pulse on its leading edge, which represents this pressure surge. This high-amplitude pressure pulse clearly has the same harmful effects as hydraulic shock. Based on the research results related to the positive opening region mentioned earlier, the variation pattern of the rear chamber oil pressure p_1 within the positive opening region is determined, in order to derive a method for selecting an appropriate positive opening size.

The simplified mathematical formulas (6.29)–(6.31) describing the variation in rear chamber oil pressure p_1 have already been listed in Chap. 6, and formulas (6.32)–(6.37), which are convenient for simulation calculations, have also been provided. The following is the calculation for state "C," representing the positive opening region, output by the simulation program in Appendix B.

$z_0 = 500$ (μm)			$V_{z0} = 0.28$(L)		$E_{zl} = 4.44$(J)	
t/μs	z_1/μm	p_1/MPa	Q_{11}/(L min^{-1})	Q_{12}/(L min^{-1})	u_p/(m s^{-1})	p/MPa
5.64	466.67	12.34	367.4	7.94	4.23	11.21
11.23	433.33	13.24	353.46	21.31	4.22	11.21
16.77	400	14.07	336.29	37.86	4.21	11.22
22.28	366.67	14.84	316.62	56.87	4.21	11.22
27.74	333.33	15.58	294.89	77.9	4.2	11.23
33.17	300	16.29	271.36	100.69	4.19	11.24
38.55	266.67	16.97	246.23	125.05	4.18	11.25
43.9	233.33	17.64	219.64	150.83	4.17	11.27
49.2	200	18.29	191.71	177.92	4.16	11.29
54.47	166.67	18.93	162.53	206.24	4.15	11.31
59.71	133.33	19.56	132.18	235.7	4.14	11.34
64.91	100	20.19	100.7	266.25	4.13	11.37
70.07	66.67	20.81	68.16	297.84	4.12	11.4
75.2	33.33	21.42	34.58	330.44	4.11	11.44

$z_0 = 600$(μm)			$V_{z0} = 0.37$(L)		$E_{zl} = 5.08$(J)	
t/μs	z_1/μm	p_1/MPa	Q_{11}/(L min^{-1})	Q_{12}/(L min^{-1})	u_p/(m s^{-1})	p/MPa
6.78	560	10.88	414.1	−5.43	4.25	11.25
13.5	520	14.36	441.68	31.66	4.24	11.24
20.16	480	11.45	364.02	12.29	4.24	11.24
26.75	440	12.09	342.94	32.75	4.23	11.26
33.29	400	12.76	320.25	54.79	4.22	11.27
39.76	360	13.42	295.58	78.77	4.21	11.28
46.18	320	14.07	268.99	104.63	4.21	11.29
52.55	280	14.7	240.63	132.24	4.2	11.3
58.86	240	15.33	210.59	161.48	4.19	11.32
65.12	200	15.94	178.98	192.25	4.18	11.34
71.33	160	16.55	145.89	224.47	4.17	11.37

(continued)

(continued)

$z_0 = 600(\mu m)$			$V_{z0} = 0.37(L)$		$E_{zl} = 5.08(J)$	
$t/\mu s$	$z_1/\mu m$	p_1/MPa	$Q_{11}/(L\ min^{-1})$	$Q_{12}/(L\ min^{-1})$	$u_p/(m\ s^{-1})$	p/MPa
77.49	120	17.16	111.4	258.06	4.16	11.4
83.6	80	17.76	75.56	292.97	4.15	11.44
89.67	40	18.36	38.41	329.15	4.13	11.48

$z_0 = 700(\mu m)$			$V_{z0} = 0.48(L)$		$E_{zl} = 5.09(J)$	
$t/\mu s$	$z_1/\mu m$	p_1/MPa	$Q_{11}/(L\ min^{-1})$	$Q_{12}/(L\ min^{-1})$	$u_p/(m\ s^{-1})$	p/MPa
7.92	653.33	8.28	421.29	⁻18.37	4.29	11.36
15.75	606.67	11.08	452.54	⁻11.04	4.28	11.35
23.49	560	14.03	470.09	51.35	4.27	11.35
31.15	513.33	11.4	388.49	9.71	4.26	11.35
38.72	466.67	11.55	355.52	22.56	4.26	11.37
46.22	420	12.03	326.54	50.84	4.25	11.38
53.65	373.33	12.58	296.76	79.88	4.24	11.39
60.99	326.67	13.14	265.44	110.44	4.23	11.4
68.27	280	13.72	232.43	142.66	4.22	11.42
75.48	233.33	14.29	197.71	176.54	4.21	11.44
82.63	186.67	14.87	161.32	212.05	4.2	11.47
89.7	140	15.44	123.31	249.15	4.19	11.5
96.72	93.33	16.02	83.73	287.79	4.18	11.54
103.67	46.67	16.6	42.62	327.93	4.17	11.59

Above are the cases for positive opening values of $z_0 = 500$, 600, and 700 μm, showing the variations of rear chamber pressure p_1, flow rates Q_{11}, Q_{12} (as shown in Fig. 4.10), piston velocity u_p, and input pressure p across the entire positive opening range, with changes in valve opening z_1 and z_2. V_{z0} and E_{zl} represent the volume of fluid drained through valve port z_1 and the corresponding energy loss in the positive opening region, respectively.

The rear chamber pressure p_1 of the three impact mechanisms with different positive openings, as obtained from simulations, is illustrated in Fig. 7.20. As z_0 increases, the peak pressure in the rear chamber decreases noticeably.

Since the flow rate $Q_1 = A_1 u_p$ and pressure p are already included in the calculations of Eqs. (6.29)–(6.31), they are essentially functions of time. Therefore, using these equations is not convenient for analysis and even less suitable for calculating the selection of the positive opening value z_0. From the output results of the 'C' state shown above, it can be observed that the piston velocity u_p and pressure p vary little

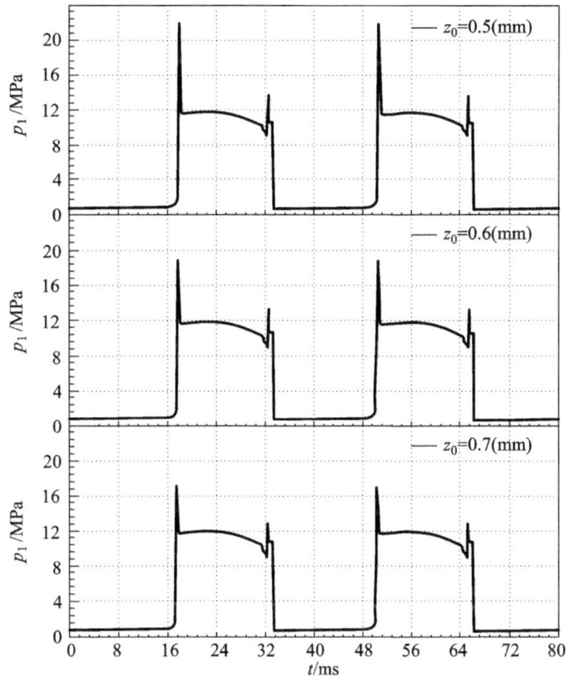

Fig. 7.20 Rear chamber pressure p_1 corresponding to different positive opening values z_0

across the entire positive opening range. In other words, the flow rate Q_1 and pressure p can be treated as constant within this range. Specifically, the maximum velocity u_{rm} during the return acceleration phase can be used to calculate the flow rate $Q_1 = A_1 u_{rm}$ in this range. The pressure p can be considered as the time-averaged pressure p_{ave} or a selected nominal pressure, with the return oil pressure p_l neglected. This approach leads to a flow rate equation that is independent of time and only depends on the opening values z_1 and z_2:

$$Q_1 = Q_{11} + Q_{12}. \tag{7.38}$$

That is

$$A_1 u_{rm} = C_7 z_1 \sqrt{p_1} + sgn(p_1 - p) z_2 C_7 \sqrt{|p_1 - p|} \tag{7.39}$$

where,

$$z_1 = z_0 - z_2 \tag{7.40}$$

The expression for p_1 derived from Eqs. (7.38)–(7.40) is similar in form to Eqs. (6.29)–(6.31), except that p_1 is solely a function of the opening values z_1 and z_2.

In Fig. 7.21, the five curves from top to bottom correspond to $z_0 = 0.3$, 0.4, 0.5, 0.6, and 0.7 mm, with the same p and u_{rm}. According to Eqs. (6.29)–(6.31), p_1 was calculated point by point by varying z_2 (z_1) within the z_0 range. The figure shows that the smaller the value of z_0, the faster p_1 decreases as z_1 increases.

By finding the extremum of Eq. (7.39), it can be determined that the rear chamber oil pressure reaches its maximum value when $z_1 = 0$(i.e., $z_2 = z_0$):

$$p_{1\max} = p + C_8\left(\frac{A_1 u_{rm}}{z_0}\right)^2 = p + C_8\left(\frac{Q_1}{z_0}\right)^2 \tag{7.41}$$

Equation (7.41) indicates a strong relationship between the maximum value of rear chamber pressure and the opening value z_0. The curve in Fig. 7.22 reflect the relationship between p_{1max} and z_0. This curve shows that when z_0 increases beyond a certain point, p_{1max} decreases slowly. Conversely, as z_0 decreases below a certain point, $p_{1\max}$ increases sharply.

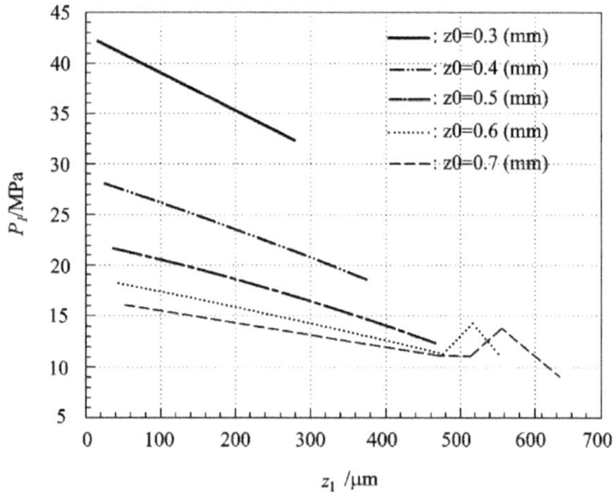

Fig. 7.21 Relationship between p_1 and z_0, z_1

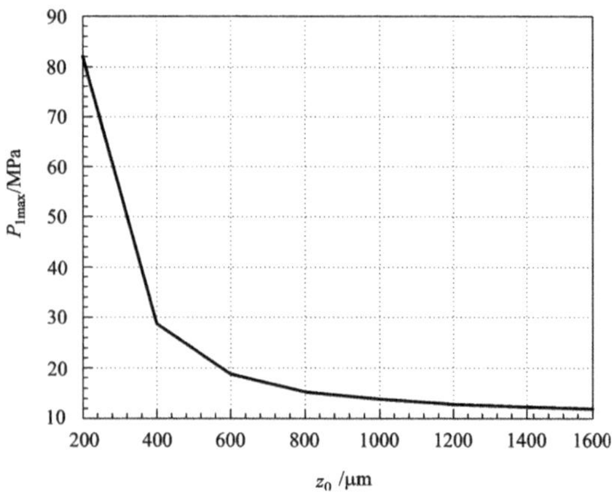

Fig. 7.22 Relationship between $p_{1\max}$ and z_0

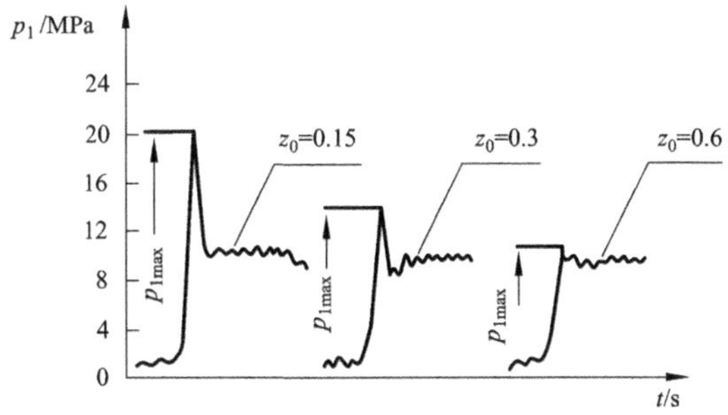

Fig. 7.23 Experimental curve of the relationship between p_{1max} and z_0

Defining a suitable range for p_{1max}, the appropriate positive opening value for a specific scheme can be determined using Eq. (7.38). Typically, we can choose $p_{1max} = 1.5p$, then there is:

$$z_0 \geq \frac{A_1 u_{rm}}{C_7\sqrt{0.5p}} \approx 15140\frac{A_1 u_{rm}}{d_v\sqrt{p}}(mm). \tag{7.42}$$

In the equation, $C_7 = c\pi d_v\sqrt{\frac{2}{\rho}}$;
$c = 0.62$;
$\rho = 870\,\text{kg/m}^3$;

d_v: valve spool diameter, which corresponds to d_4 in the simulation program. All input values for the calculation use the International System of Units (SI).

The working principle of the YYG90 hydraulic rock drill impact mechanism is of the type shown in Fig. 7.18. In Fig. 7.23, from left to right, the waveforms represent the measured oil pressure variations in the rear chamber of the piston for z_0= 0.15, 0.3, and 0.6 mm. The pulse peak values p_{1max} are close to the calculated values.

Referring to Fig. 7.19, it can be seen that the dual-control hydraulic impact mechanism can also select the z_0 value according to the Eq. (7.42).

2. Oil Pressure Surge at the End of the Impact Phase

In the rear-control hydraulic impact mechanism shown in Fig. 7.18, the front chamber operates at constant pressure and is not controlled by the valve port, thus avoiding the issue of oil pressure surges caused by the valve port narrowing. In the dual-control hydraulic impact mechanism shown in Fig. 7.19, during the end of the impact phase, both the piston and valve spool move to the left. When the piston impacts the end of the rod at a speed of u_{im}, if the valve port z_4 controlling the front chamber is too small, a significant oil pressure surge can occur in the front chamber. If at this time p_2 is restricted to $p_2 = 1.5p$, then:

$$A_2 u_{im} \leq C_7 z_4\sqrt{1.5p}$$

$$z_4 \geq \frac{A_2 u_{im}}{C_7\sqrt{1.5p}} \approx 8740\frac{A_2 u_{im}}{d_v\sqrt{p}}(mm) \tag{7.43}$$

The units used in the equation are the same as those in Eq. (7.42).

If the position of the impact stroke reversal feedback port S_{ic} is determined by the requirement that "when the piston impacts the end of the drill rod, the valve spool is still a distance z_4 away from the neutral position," the oil pressure surge in the front chamber can be limited to the specified range. Would such an impact lead to reversing lag (i.e., the piston starts its rebound motion, but the valve spool has not yet passed the neutral position)? Theoretically, the answer is no. After the piston impacts the end of the drill rod, there is a brief pause of approximately 0.5μs, during which the valve can fully pass the neutral position without causing any lag.

3. **Cavitation at the End of the Return Stroke Acceleration Phase**

In rear-control impact mechanisms, cavitation does not occur at the end of the return stroke acceleration phase due to the narrowing of the valve port. However, in dual-control impact mechanisms, cavitation does occur during this phase (see Fig. 7.19). At the end of the return stroke acceleration phase, the piston speed is high, increasing the demand for flow rate in the front chamber. Clearly, the valve spool, which is undergoing a reversing motion, gradually reduces the size of the valve port z_3 controlling the front chamber. When z_3 becomes sufficiently small, even if the front chamber oil pressure $p_2 = 0$, the flow rate through z_3 can no longer meet the front chamber's flow demand (i.e., $C_7 z_3\sqrt{p} \leq A_2 u_{rm}$), resulting in cavitation in the front chamber. The size of the cavitated volume ΔV_c can be expressed by Eq. (7.44):

$$\Delta V_c = \int_{t_1}^{t_2}\left(A_2 u_p - C_7 z_3\sqrt{p - p_2} - C_7 z_4\sqrt{p_a - p_l}\right)dt. \tag{7.44}$$

where t_1: the moment when $p_2 = 0$;

t_2: the moment when z_4 is large enough to meet the front chamber's oil flow demand;

p_a: atmospheric pressure or return oil pressure.

If we set $u_p = u_{rm} = \text{Const}$; $p_2 = 0$; $p_a = 0.1$ MPa, and $p_l = 0$, an approximate solution can be obtained as:

$$\Delta V_c \approx \frac{A_2^2 u_{rm}^2}{C_7 V_v}\cdot\left(1 + \frac{1}{\sqrt{p}} - \frac{1}{2C_7} - \frac{1}{2C_7 p}\right) \tag{7.45}$$

The cavitation formed at this time in the impact mechanism of YYG80 hydraulic rock drill is approximately 0.6 cm^3, as calculated using Eq. (7.45).

Is it possible to prevent this harmful phenomenon by increasing the positive opening value z_4 and supplying oil from the return path? Due to the low return oil pressure, a larger value of z_4 (up to over 1 mm) would be required, which is clearly not suitable.

4. **Cavitation at the End of the Impact stroke Phase**

Referring to Figs. 7.18 and 7.19, similarly, it can be concluded that when the piston impacts the end of the rod, the size of the control ports in this two types impact mechanisms must ensure

$$z_2 \geq 15140\frac{A_1 u_{im}}{d_v \sqrt{p}}(\text{mm}) \tag{7.46}$$

in order to prevent cavitation in the rear chamber of the piston. When determining the position of the impact stroke reversing feedback port S_{ic}, one must consider the possibility of cavitation due to improper selection of z_2.

5. **Pressure Surge in Rear Chamber Pressure During the Impact Moment**

At the moment when the piston impacts the target object (such as the drill tail or rod), a pressure surge may also occur in the rear chamber. The measured rear chamber waveform, approximating a square wave, shown in Fig 7.17, displays a pulse on the trailing edge, signifying this type of pressure surge. There are two main causes for this pressure surge.

(1) At the moment of piston impact, the valve spool's stroke reversal exhibits a slight delay. That is, at the end of the aforementioned 'G' state, the valve has not yet crossed the neutral position ($y_v < S_{1v}$), and the rear chamber remains in a high- pressure state. At this point, as the piston's impact velocity drops abruptly from its maximum to zero, the velocity of the working medium in the entire inlet pipe also drops suddenly to zero, generating a typical hydraulic shock phenomenon (similar to a sudden valve closure). According to relevant data, the maximum pressure surge caused by this shock can be calculated using the following formula:

$$\Delta p = a\rho u$$

where:

a: transmission velocity of the pressure wave in the pipe, here take $a = 1200\,\text{m/s}$;

ρ: density of the medium(kg/m^3);

u: the velocity of the medium before impact (m/s).

The pressure amplitude of the hydraulic shock generated in the rear chamber at the moment of impact can be approximately calculated using this formula as:

$$p_1 = \hat{p}_1 + a\rho u_{im} \tag{7.47}$$

$\hat{p}_1$ is the rear chamber pressure from the previous step, just before the moment of impact.

Since the rise of the hydraulic shock takes a certain amount of time, it can be handled simply in the simulation. It can be assumed that before the valve crosses the neutral position, the rear chamber pressure remains at the level indicated by Eq. (7.47).

(2) Significant delay in valve stroke reversal. When the piston finishes its impact and pause and begins to rebound, the valve still has not crossed the neutral position and remains in the abnormal 'A″' state. The pressure surge situation is similar to the one described above, but due to the longer duration, the pressure rise caused by hydraulic shock is more complete, with a higher amplitude. However, if the valve is near the neutral position (in the positive opening range) when the piston starts rebounding, the situation changes. In this case, the rear chamber is almost in a closed state, and the rebound energy of the piston is converted into pressure energy, causing a sharp rise in the rear chamber pressure. The rebound energy is wasted, unlike in the previous situation, where some of the increased pressure energy could be stored by the accumulator (or as medium compression). Clearly, calculating this process is quite complex. Since this phenomenon is highly random, it will not be analyzed in detail in this book. The pressure surge will still be calculated using Eq. (7.47).

Through the above discussion on the harmful phenomena of pressure surges and cavitation in hydraulic impact mechanisms, our understanding of their working principles has been further deepened. We can also use the calculation formulas provided in this book to select the appropriate positive opening value or stroke reversing position to mitigate and eliminate these harmful phenomena. Moreover, the analysis indicates that, in terms of overcoming the two harmful effects of pressure surges and cavitation, the rear-control hydraulic impact mechanism is superior to the dual-control impact mechanism.

Appendix A
Calculation Program of Hydraulic Impact Mechanism by Three-Step Analysis

This program is written in Python. Relevant explanations are included after the corresponding program statements. For the meanings of the variable codes representing the structural parameters of the piston and valve spool, please refer to Fig. 4.1. In the program's input parameters, the units are as follows: length in meters (m), pressure in megapascals (MPa), velocity in meters per second squared (m/s^2), and frequency in hertz (Hz).

1. The variable symbols in the program correspond to the symbols used in the formulas from Chaps. 1 to 3 as follows:

   EI: Ei, impact energy (J); F: f, impact frequency (Hz);
   UIM: u_{im}, impact velocity; PIN: working pressure including local loss (MPa);
   NU: μ, dynamic viscosity coefficient (m^2/s);
   EPSI: ε, eccentricity ratio;
   LAM: λ, ratio of the maximum to minimum operating pressure of the accumulator, $\lambda = 1.25–1.4$;

RHO: ρ, oil density (kg/m^3);	BETA: β;
BE1: β_1	BE2: β_2
PSISP: ψ_{sp}	PSIUR: ψ_{ur}
PSIA1: ψ_{A1}	PSIA2: ψ_{A2}
PHIL: φ_l	PHIC: φ_c
PHIO: φ_o	PHIY: φ_y
PHIP: φ_p	PHIQ: φ_q
PHIU: φ_u	ETV: η_v
ETA: η	ETM: η_m
ETP: η_p	DEP: Δp
DVRI: ΔV_{ri}	DVRO: ΔV_{ro}
DVI: ΔV_i	DVO: ΔV_o

Q. He, *Research and Design of Hydraulic Impact Mechanisms*,
https://doi.org/10.1007/978-981-96-4864-1

2. In the output results, the output symbols correspond to the symbols used in the formulas from Chaps. 1 to 3 as follows:

Be1: β_1	Be2: β_2
PsiSp: ψ_{sp}	PsiUr: ψ_{ur}
PsiA1: ψ_{A1}	PsiA2: ψ_{A2}
PhiY: φ_y	PhiO: φ_o
PhiU: φ_u	PhiC: φ_c
PhiL: φ_l	PhiP: φ_p
PhiQ: φ_q	

Eta, Etv, Etm, and Etp: represent η, η_v, η_m, and η_p respectively;
DVri, Dvro, DVi, and DVo: represent ΔV_{ri}, ΔV_{ro}, ΔV_i, and ΔV_o respectively.

3. Program code:

```
import math
import numpy as np

KY = 0.08 # comprehensive resistance coefficient(0.06-0.12)
KO = 0.10 # return oil resistance coefficient(0.05~0.12)
ZETA = 300 #  local resistance coefficient(180~400)
NU = 0.4 * 1e-4 # viscosity(m^2/s)
EPSI = 0.95 # ring eccentricity(0.9~1.0)
RHO = 870 # oil density
LAM = 1.35
# ratio of the maximum to minimum operating pressure of the
# accumulator(1.25-1.4)

D8 = 0.025
D = 0.062
H = 50 * 1e-6
L = 0.04
L3 = 0.045
H1 = 30 * 1e-6
L1 = 0.05
H2 = 25 * 1e-6
L2 = 0.05
PII = 3.14159
KY1 = 1 - KY
KY2 = 1 + KY

def main_sub():
     if AAA == 2: # if AAA=2 represents dual-control type
          BE1 = (KY1 ** 2 - KO ** 2) / (KY1 * KY2 + 2 * KO * KY
                                            * BETA - KO ** 2)
          BE2 = (KY1 * BETA + KO) / (KY1 + KO * BETA)
     else: # rear-control type
          BE1 = (KY1 ** 2 - KO * KY2) / (KY1 * (KY2 - KO)
                                              + 4 * KY * BETA)
          BE2 = (KY1 * BETA + KO) / (KY1 + KY2 * BETA)
```

```
    PSISP = BETA / 2 / (BETA + np.sqrt(BETA * (1 + BETA * BE1)))
    PSIUR = np.sqrt(BETA / (1 + BETA * BE1))
    SP = PSISP * UIM * T # piston stroke
    SD = BE1 * PSIUR ** 2 * SP # return deceleration stroke
    SR = PSIUR ** 2 / BETA * SP # return acceleration stroke
    TR = 2 * PSISP * PSIUR / BETA * T # return acceleration time
    TD = 2 * BE1 * PSIUR * PSISP * T # return deceleration time
    TI = 2 * PSISP * T # impact stroke time
    WI = UIM / T / PSISP / 2 # impact stroke acceleration
    WR = BETA * WI # return acceleration phase acceleration
    WD = WI / BE1 # return deceleration phase acceleration
    URM = PSIUR * UIM  #maximum velocity during return
                  # acceleration phase

  if AAA == 2: # AAA=2 represent dual-control type
    PSIA1 = (KY1 + KO * BETA) / (KY1 ** 2 - KO ** 2) / PSISP / 2
    PSIA2 = (KO + KY1 * BETA) / (KY1 ** 2 - KO ** 2) / PSISP / 2
    K1 = 4 * RHO * ZETA * PSIA1 * PSISP / BETA / A8 / A8 / 3
    K2 = 2 * PSIA1 * PSISP
    PHIL = 2 * PSISP * (K2L * PSIUR / BETA + K1L * (1 + BE1 * PSIUR))
    PHIU = 2 * PSIA1 * PSISP * PSIUR ** 2 * (BE2 / BETA - BE1)
    PHIY = 2 * KY * (1 + BE2 + BE1 * (1 + BE2) * PSIUR ** 2)\
         * PSIA1 * PSISP
    PHIO = 2 * KO * PSIA1 * PSISP * (1 + BE2 - BE1 * PSIUR ** 2)
    PHIC = K1 * (BE2 ** 2 * PSIUR ** 3 + BETA * (1 + BE1 * PSIUR ** 3))
    PHIQ = K2 * (1 + BE2) * PSIUR ** 2 / BETA
    PHIP = 2 * (1 + BE2) * PSIUR ** 2 * PSIA1 * PSISP * EI\
         / BETA / T / PI / np.sqrt(PI)
    ETV = K2 * EI * BETA / (K2 * EI * (1 + BE2) * PSIUR ** 2 + PHIL *
                       BETA * T * PI ** 2)
   # volumetric efficiency
    DVRI = 2 * (1 + BE2) ** 2 * PSIA1 * PSIUR ** 4 * PSISP ** 3 / BE2\
    / BETA ** 3 * EI / PI
    DVRO = 2 * BE2 * PSIA1 * PSIUR ** 2 * PSISP / BETA\
    * (1 - (1 + BE2) * PSIUR * PSISP / BE2 / BETA) ** 2 * EI / PI
    DVI = (2 * PSIA1 * PSIUR ** 2 * PSISP * (BE1 + 2 * BE1 * (1 + BE2)
    * PSIUR * PSISP  / BETA + ((1 + BE2) * PSIUR * PSISP / BETA) ** 2)
    * EI / PI); DVO = 2 * PSIA1 * PSISP * (1 - (1 + BE2) * PSIUR ** 2 *
                               PSISP / BETA) ** 2 * EI / PI
    if BE2 <= (1 + BE2) * PSIUR * PSISP / BETA:
      DVRI = 0
      DVRO = 0
      DVI = DVO
  else:  #AAA=1 represents rear-control type
    PSIA1 = (KY1 + KY2 * BETA) / (KY1 ** 2 - KO * KY2) / PSISP / 2
    PSIA2 = (KO + KY1 * BETA) / (KY1 ** 2 - KO * KY2) / PSISP / 2
    K1 = 4 * RHO * ZETA * PSIA1 * PSISP / BETA / A8 / A8 / 3
    K2 = 2 * PSIA1 * PSISP
    PHIL = K2L + 2 * K1L * PSISP * (1 + BE1 * PSIUR)
    PHIU = 2 * PSIA1 * PSISP * (BE2 - BE1 * PSIUR ** 2)
    PHIY = 2 * KY * (1 + 2 * BE2 + BE1 * PSIUR ** 2) * PSIA1 * PSISP
    PHIO = 2 * KO * PSIA1 * PSISP * (1 - BE1 * PSIUR ** 2)
    PHIC = K1 * (BE2 ** 2 * PSIUR ** 3 + BETA * (1 - BE2) **
            2 * (1 + BE1 * PSIUR ** 3))
```

```
        PHIQ = K2 * PSIUR ** 2 / BETA
        PHIP = 2 * PSIUR ** 2 * PSIA1 * PSISP * EI / BETA / T / PI\
            / np.sqrt(PI)
        ETV = K2 * EI * BETA * (1 -BE2) / (K2 * EI * PSIUR ** 2 +
                                    PHIL * BETA * T * PI ** 2)
        # volumetric efficiency
        DVRI = 2 * PSIA1 * PSIUR ** 4 * PSISP ** 3 / BE2 /\
             BETA ** 3 * EI / PI
        DVRO = 2 * BE2 * PSIA1 * PSIUR ** 2 * PSISP / BETA * (1 - PSIUR *
        PSISP / BE2 / BETA) ** 2 * EI / PI
        DVI = 2 * PSIA1 * PSIUR ** 2 * PSISP * ((1 - BE2)
        * BE1 + 2 * BE1 * PSIUR * PSISP / BETA + (PSIUR * PSISP
        / BETA) ** 2 / (1 - BE2)) * EI / PI
        DVO = 2 * (1 - BE2) * PSIA1 * PSISP * (1 - PSIUR ** 2 * PSISP
        / BETA / (1 - BE2)) ** 2 * EI / PI
        if BE2 <= PSIUR * PSISP / BETA:
            DVRI = 0
            DVRO = 0
            DVI = DVO

    if DVO > DVI:
        DV = DVO
    else:
        DV = DVI
    VA = DV * (LAM ** (1 / 1.4) + 1) / (LAM ** (1 / 1.4) - 1)
    PA = PI / 2 ** 1.4
    resilt_dict = {"DVRO": DVRO, "VA": VA, "PA": PA, "PSIA1": PSIA1,
    "PSIA2": PSIA2, "PHIL": PHIL, "PHIU": PHIU, "PSISP": PSISP,
    "PHIY": PHIY, "PHIO": PHIO, "PHIC": PHIC, "PHIQ": PHIQ,
    "DV": DV, "PHIP": PHIP, "ETV": ETV, "DVRI": DVRI, "BE1":
     BE1, "BE2": BE2, "SD": SD, "SR": SR, "TR": TR, "TD": TD,
     "TI": TI, "WR": WR, "WD": WD, "URM": URM, "SP": SP,
    "WI": WI, "DVI": DVI, "DVO": DVO, "PSIUR": PSIUR}
    return resilt_dict

def format_value(value):
    if abs(value) >= 1e3 or 0 < abs(value) < 1e-3:
        return "{:8.3e}".format(value)
    else:
        return "{:8.3f}".format(value)

if __name__ == '__main__':
    AAA, EI, F, UIM, PIN = map(float, input
    ("Please enter AAA(rear-control:1,dual-control:2),Ei,f,Uim,"
     "PiN(Separated by spaces):").split())
    MP = 2 * EI / UIM ** 2
    A8 = PII * D8 * D8 / 4
    MU = RHO * NU
    KL = PII * (1 + 2.5 * EPSI * EPSI) / 12 / MU
    D1 = 0
    D2 = 0
    K1L = KL * (D * H ** 3 / L3 + D1 * H1 ** 3 / L1)
    K2L = KL * (D * H ** 3 / L + D2 * H2 ** 3 / L2)

    if AAA == 1:
```

```
    print("Rear-Control")
  elif AAA == 2:
    print("Dual-Control")
  T = 1 / F

  M = 1000000
  W = 60000
  S = 1000
  PIN = M * PIN
  PI = PIN
  BB = 0
# The for loop calculates the two characteristic values
#of the accumulator.
  for i in range(200, 801):
    BETA = 0.001 * i
    if AAA == 2:
      BE1 = (KY1 ** 2 - KO ** 2) / (KY1 * KY2 + 2 * KO
      * KY * BETA - KO ** 2)
      BE2 = (KY1 * BETA + KO) / (KY1 + KO * BETA)
    else:
      BE1 = (KY1 ** 2 - KO * KY2) / (KY1 * (KY2 - KO) + 4 * KY * BETA)
      BE2 = (KY1 * BETA + KO) / (KY1 + KY2 * BETA)
    PSISP = BETA / 2 / (BETA + np.sqrt(BETA * (1 + BETA * BE1)))
    PSIUR = np.sqrt(BETA / (1 + BETA * BE1))
    US = PSIUR * PSISP / BETA
    PUS = 2 * PSIUR ** 2 * PSISP
    BU1 = BE1 * PSIUR + 1
    BU2 = BE1 * PSIUR ** 2 - 1
    if AAA == 2:
      if BB > 0:
        if (1 + BE2) * PUS * BU1 + BETA * BU2 > 0:
          BET2 = BETA - 0.0005 # β^(**)
          break
      else:
        if BE2 / (1 + BE2) - US > 0:
          BET1 = BETA - 0.0005 # β^*
          BB = 1
    else:
      if BB > 0:
        if PUS * BU1 + BETA * (1 - BE2) * BU2 > 0:
          BET2 = BETA - 0.0005
          break
      else:
        if BE2 - US > 0:
          BET1 = BETA - 0.0005
          BB = 1

  print("Beta * ={:.4f}\tBeta * * ={:.4f}".format(BET1, BET2))
# Output the accumulator's return oil discharge volume,
# design discharge volume, and inflation volume corresponding
# to the return-impact acceleration ratio Beta between the
# two characteristic values. Based on these data,
# select an appropriate Beta value for input.
  DBET = (BET2 - BET1) / 10
  BETA = BET1
```

```
BET = np.zeros(10)
Dvro = np.zeros(10)
Dv = np.zeros(10)
Va = np.zeros(10)
for i in range(0, 10):
   BETA = BETA + DBET
   BET[i] = BETA
   dict1 = main_sub()
   Dvro[i] = dict1["DVRO"]
   Dv[i] = dict1["DV"]
   Va[i] = dict1["VA"]
print("Beta\t\tDVro\t\tDV\t\t\tVa")

for i in range(0, 10):
   print("{:.4f}\t\t{:.4f}\t\t{:<5.4f}\t\t{:.4f}"
   .format(BET[i], M * M * Dvro[i], M * M
   * Dv[i], M * M * Va[i]))
print("Please Select Beta (select a Beta value and input it)")
BETA = float(input("Beta="))
dict2 = main_sub()
PHIL = dict2["PHIL"]
PHIQ = dict2["PHIQ"]
PHIU = dict2["PHIU"]
PHIC = dict2["PHIC"]
PHIO = dict2["PHIO"]
PHIY = dict2["PHIY"]
PHIP = dict2["PHIP"]

QI = PHIL * PI + PHIQ * EI / PI / T
EC = PHIC * QI * EI ** 2 / PI ** 2 / T  #local lost energy
DEP = EC / QI / T # local pressure loss
PI = PIN - DEP
print("DEP:{:<8s}MPa\tPI:{:<8s}MPa\t"
    .format(format_value(DEP / M), format_value(PI / M)))
dict3 = main_sub()
PSIA1 = dict3["PSIA1"]
PSIA2 = dict3["PSIA2"]
PSIUR = dict3["PSIUR"]
PA = dict3["PA"]
PHIP = dict3["PHIP"]
PSISP = dict3["PSISP"]
TR = dict3["TR"]
TD = dict3["TD"]
TI = dict3["TI"]
SR = dict3["SR"]
SD = dict3["SD"]
SP = dict3["SP"]
WR = dict3["WR"]
WI = dict3["WI"]
WD = dict3["WD"]
URM = dict3["URM"]
ETV = dict3["ETV"]
DVRI = dict3["DVRI"]
DVI = dict3["DVI"]
DVO = dict3["DVO"]
```

```
DVRO = dict3["DVRO"]
VA = dict3["VA"]
BE1 = dict3["BE1"]
BE2 = dict3["BE2"]

EC = PHIC * QI * EI ** 2 / PI ** 2 / T #local energy loss
EL = PHIL * T * PI ** 2 # leakage loss energy
EU = PHIU * EI # stroke loss energy
EO = PHIO * EI # return oil resistance energy loss
EY = PHIY * EI # comprehensive resistance energy loss
EIN = EI + EC + EL + EY + EO #Total input energy
ETM = 1 - (EO + EY) / EIN # mechanical efficiency
ETP = 1 - EC / EIN # pressure efficiency
ETA = EI / EIN # overall efficiency

QI = PHIL * PI + PHIQ * EI / PI / T #Calculate the input flow rate
A1 = PSIA1 * MP * UIM / PI / T
A2 = PSIA2 * MP * UIM / PI / T
D1 = np.sqrt(D * D - 4 * A1 / PII)
D2 = np.sqrt(D * D - 4 * A2 / PII)

print("EL:{:<8s} J\tEC:{:<8s} J\tEO:{:<8s} J\tEIN:{:<8s} J"
.format(format_value(EL), format_value(EC),
format_value(EO), format_value(EIN)))
        print("Ei:{:<8s}  J\tf:{:<8s}  Hz\tUim:{:<8s}  m/
s\tPin:{:<8s} MPa"
.format(format_value(EI), format_value(F), format_value(UIM),
format_value(PIN / M)))
print("K1l:{:<8s} \tK2l:{:<8s} ".format(format_value(K1L),
 format_value(K2L)))
print("Ky:{:<8s} \tKo:{:<8s} \tZeta:{:<8s}"
.format(format_value(KY), format_value(KO),
format_value(ZETA)))
print("Beta:{:<8s}\tBe1:{:<8s}\tBe2:{:<8s}"
.format(format_value(BETA), format_value(BE1),
format_value(BE2)))
print("PsiSP:{:<8s}\tPsiUr:{:<8s}\tPsiA1:{:<8s}"
"\tPsiA2:{:<8s}".format(format_value(PSISP),
        format_value(PSIUR),  format_value(PSIA1),  format_
value(PSIA2)))
print("PhiY:{:<8s}\tPhiO:{:<8s}\tPhiU:{:<8s}\tPhiC:{:<8s}"
.format(format_value(PHIY), format_value(PHIO),
format_value(PHIU), format_value(PHIC)))
print("Phil:{:<8s}\tPhiP:{:<8s}\tPhiQ:{:<8s}"
.format(format_value(PHIL), format_value(PHIP)
, format_value(PHIQ)))
print("tr:{:<8s} ms\ttd:{:<8s} ms\tti:{:<8s} ms"
.format(format_value(S * TR), format_value(S * TD),
format_value(S * TI)))
print("Sr:{:<8s} mm\tSd:{:<8s} mm\tSp:{:<8s} mm"
.format(format_value(S * SR), format_value(S * SD),
format_value(S * SP)))
     print("wr:{:<8s}  m/s^2\twi:{:<8s}  m/s^2\twd:{:<8s}  m/
s^2\tUrm:{:<8s} m/s"
.format(format_value(WR), format_value(WI), format_value(WD),
```

```
 format_value(URM)))
print("A2:{:<8s} cm^2\tA1:{:<8s} cm^2\tMp:{:<8s} kg"
.format(format_value(10000 * A2), format_value(10000 * A1)
, format_value(MP)))
print("d1:{:<8s} mm\td2:{:<8s} mm"
.format(format_value(D1 * 1000), format_value(D2 * 1000)))
print("Qi:{:<8s} L/min\tpi:{:<8s} MPa\tDep:{:<8s} MPa"
.format(format_value(QI * W), format_value(PI / M),
format_value(DEP / M)))
print("Eta:{:<8s}\tEtv:{:<8s}\tEtm:{:<8s}\tEtp:{:<8s}"
.format(format_value(ETA), format_value(ETV),
format_value(ETM), format_value(ETP)))
print("DVri:{:<8s}\tDVro:{:<8s}"
.format(format_value(DVRI), format_value(DVRO)))
print("DVi:{:<8s}\tDVo:{:<8s}"
.format(format_value(DVI), format_value(DVO)))
print("Va:{:<8s} ml\tpa:{:<8s} MPa"
.format(format_value(M * VA),
format_value(PA / M)))
```

Appendix B
Simulation Program for Hydraulic Impact Mechanisms

This program is written in Python. It is primarily designed for computer simulations of multi-state and multi-condition hydraulic impact mechanisms, similar to the YYG250 rear-control type described in the book.

1. Components of the Program

(1) Parameter Setting Table

The parameter initialization input is implemented through the parameter setting table (.xlsx file), which includes the structural parameters of the hydraulic impact mechanism and other system parameters (such as oil viscosity coefficient and oil density).

(2) Parameter Assignment

After main program runs, the paramaters.py program will be automatically loaded, and parameter will be assigned by paramaters.py.

(3) Main Program

The main program of this software adopts a modular design method.

Firstly, create an Excel spreadsheet of the specific structural and system parameters of the hydraulic impact mechanism.

Next, place parameters.py and the main program in the same directory, run the main program in the Python environment. In the terminal, a prompt will appear: **"Please enter the path and name of the parameter file (.xlsx):"**. After entering the parameter file name and its path, the program will start simulation calculation.

Upon completion, the simulation results will be output. The simulation results include: impact frequency, impact energy, final impact velocity, stroke, various efficiencies, and more. The output will also include simulation curves. Follow the menu prompts to display and save the desired curves. The selected curves will be saved in the same directory as the main program.

Q. He, *Research and Design of Hydraulic Impact Mechanisms*,
https://doi.org/10.1007/978-981-96-4864-1

2. The Meaning of the Main Codes in the Program

The codes of the input parameters and the corresponding units used:

QI: inlet flow (L/min);

PA, PLA: Inflation pressure of high and low pressure accumulators (MPa);

K: Rebound coefficient after piston impact;

TT: Simulation calculation time step (s);

DTV: The difference in chamber volume of the high-pressure accumulator at the end of the first and last cycles in simulation calculation (ml), as a criterion for the completion of simulation calculations;

TAUP, TAUV: the coefficient to calculate hydraulic clamping force when the piston and valve core move;

MU: the kinematic viscosity of the medium (m^2/s);

EPSP, EPSV: Eccentricity of piston and valve core axis relative to its supporting shaft sleeve axis during movement;

RHO: Density of working medium (Kg/m^3);

ZT1–ZT6: Local resistance coefficient (−), as shown in Fig. 4.4;

C: Valve port flow coefficient;

D, H, L, H1, L1…D3, H3, L3…: The diameters of the mating cylindrical surfaces of the piston and valve core (m), clearance (m) and length (m). Please refer to Fig. 4.1;

MP, MV: mass of piston and valve core (Kg);

TST: The pause time after piston impact (s);

D0, L0: Inner Diameter of Return Pipe (M) and Length (M);

DH, LH: The inner diameter (m) and length (m) of the high-pressure oil inlet hose are calculated based on only the section near the impact mechanism, as the length effect of the hose is ignored;

D7, D8: Inner Diameter of Valve Directional Control Oil Passage (M), Design Nominal Diameter of Impact Mechanism (M);

VA, VLA: Inflation volume of high-accumulator and low-accumulator (ml);

V20: Volume of Oil Inside the Impactor (ml);

Sc, Sic: Distance of control oil passage for valve core return direction and stroke direction (mm);

SV, Z0: The stroke of the valve core and the positive opening amount at its center position;

KL, KC: Bulk elastic modulus of medium and high-pressure rubber hose (MPa);

AA: Transmission velocity of pressure waves in the medium (m/s);

LAMD: Resistance coefficient λ of medium flow in the return pipe;

QD: Rated flow rate of return oil filter;

PD: Rated pressure drop of return oil filter.

The above 60 parameters is listed in an Excel, with variable codes in the first column and parameter values in the second column. These codes must be consistent with those in parameters.py; otherwise, the program will encounter errors during execution.

3. Main Program Code

```
import numpy as np
import matplotlib.pyplot as plt
import matplotlib.image as mpimg
import paramaters as pm

def PlotDemo1(L1, L2):
  global P1_list, QE_list, P_list, QPV_list, PL_list, QE2_list, \
  QT_list,QPV_list_ni
  fig = plt.figure(figsize = (12, 8))
  ax = fig.add_subplot(1, 1, 1)
  color = 'tab:blue'
  ax.plot(L1, L2, color=color)
  if L2 == P1_list:
    ax.set_title("Rear Chamber Pressure of Piston P1")
    ax.set_xlabel('Time (s)')
    ax.set_ylabel('Rear chamber pressure pressure (MPa)')
    plt.savefig("rear chamber pressure P1.png")
  elif L2 == QE_list:
    ax.set_title("Compensation Flow Rate Qe")
    ax.set_xlabel('Time (s)')
    ax.set_ylabel('Compensation flow rate Qe (L/min)')
    plt.savefig("Compensation flow rate Qe.png")
  elif L2 == QPV_list_ni:
    ax.set_title("Flow Rate Required for Piston Valve Qpv")
    ax.set_xlabel('Time (s)')
    ax.set_ylabel('Flow rate required Qpv (L/min)')
    plt.savefig("Flow rate required Qpv.png")
  elif L2 == P_list:
    ax.set_title("Inlet Pressure P")
    ax.set_xlabel('Time (s)')
    ax.set_ylabel('Inlet pressure P (MPa)')
    plt.savefig("Inlet pressure.png")
  elif L2 == UP_list:
    ax.set_title("Velocity of Piston Up")
    ax.set_xlabel('Time (s)')
    ax.set_ylabel('Velocity of piston Up (m/s)')
    plt.savefig("Velocity of piston.png")
  elif L2 == SP_list:
    ax.set_title("Displacement of Piston Sp")
    ax.set_xlabel('Time (s)')
    ax.set_ylabel('Displacement Sp(mm)')
    plt.savefig("Displacement Sp.png")
  elif L2 == PL_list:
    ax.set_title("Return Pressure PL")
    ax.set_xlabel('Time (s)')
    ax.set_ylabel('Return pressure PL (MPa)')
    plt.savefig("Return pressure PL.png")
  elif L2 == QE2_list:
    ax.set_title("Compensation Flow Rate Qe2")
    ax.set_xlabel('Time (s)')
    ax.set_ylabel('Compensation flow rate Qe2 (L/min)')
    plt.savefig("Compensation flow rate Qe2.png")
  elif L2 == QT_list:
```

```
        ax.set_title("Inlet Flow Rate Qt")
        ax.set_xlabel('Time (s)')
        ax.set_ylabel('Inlet flow rate Qt (L/min)')
        plt.savefig("Inlet flow rate Qt.png")
    plt.show()

def PlotDemo_SP_UP_together():
    figure, ax1 = plt.subplots(figsize=(12, 8))
    ax1.set_title("Velocity and Displacement of Piston")
    color = 'tab:red'
    ax1.set_xlabel('Time (s)')
    ax1.set_ylabel('Velocity (m/s)', color=color)
    ax1.plot(T_list, UP_list, color=color)
    ax1.tick_params(axis='y', labelcolor=color)

    ax2 = ax1.twinx()
    color = 'tab:blue'
    ax2.set_ylabel('Displacement (mm)', color=color)
    ax2.plot(T_list, SP_list, color=color)
    ax2.tick_params(axis='y', labelcolor=color)
    figure.tight_layout()
    plt.savefig("Velocity_and_Displacement_together.png")
    plt.show()

def PlotDemo_Stable(L1, L2):
    # Display and save the curve plot after the simulation stabilizes
    global P1_list_stable2, P_list_stable2, QE_list_stable2, \
    QPV_list_stable2, SP_list_stable2, UP_list_stable2,\
    QE2_list_stable2,PL_list_stable2, QT_list_stable2
    fig = plt.figure(figsize=(12, 8))
    ax = fig.add_subplot(1, 1, 1)
    color = 'tab:blue'
    ax.plot(L1, L2, color=color)
    if L2 == P1_list_stable2:
        ax.set_title("Rear Chamber Pressure of Piston P1")
        ax.set_xlabel('Time (s)')
        ax.set_ylabel('Rear chamber pressure P1 (MPa)')
        plt.savefig("Stabilized Rear Chamber Pressure P1.png")
    elif L2 == QE_list_stable2:
        ax.set_title("Compensation Flow Rate Qe")
        ax.set_xlabel('Time (s)')
        ax.set_ylabel('Compensation flow rate Qe (L/min)')
        plt.savefig("Stabilized Compensation Flow Rate Qe.png")
    elif L2 == QPV_list_stable2:
        ax.set_title("Flow Rate Required for Piston Valve Qpv")
        ax.set_xlabel('Time (s)')
        ax.set_ylabel('Flow rate required Qpv (L/min)')
        plt.savefig("Stabilized Flow Rate Required Qpv.png")
    elif L2 == P_list_stable2:
        ax.set_title("Inlet Pressure P")
        ax.set_xlabel('Time (s)')
        ax.set_ylabel('Inlet pressure P (MPa)')
        plt.savefig("Stabilized Inlet Pressure P.png")
    elif L2 == UP_list_stable2:
        ax.set_title("Velocity of Piston Up")
```

```
        ax.set_xlabel('Time (s)')
        ax.set_ylabel('Velocity (m/s)')
        plt.savefig("Stabilized Velocity of Piston Up.png")
    elif L2 == SP_list_stable2:
        ax.set_title("Displacement of Piston Sp")
        ax.set_xlabel('Time (s)')
        ax.set_ylabel('Displacement of piston Sp(mm)')
        plt.savefig("Stabilized Displacement of Piston Sp.png")
    elif L2 == QE2_list_stable2:
        ax.set_title("Compensation Flow Rate QE2")
        ax.set_xlabel('Time (s)')
        ax.set_ylabel('QE2 Flow (L/min)')
        plt.savefig("Stabilized Compensation Flow Rate QE2.png")
    elif L2==QT_list_stable2:
        ax.set_title("Inlet Flow Rate Qt")
        ax.set_xlabel('Time (s)')
        ax.set_ylabel('inlet flow rate Qt (L/min)')
        plt.savefig("Stabilized Inlet Flow Rate Qt.png")
    elif L2==PL_list_stable2:
        ax.set_title("return pressure PL")
        ax.set_xlabel('Time (s)')
        ax.set_ylabel('return pressure PL (PMa)')
        plt.savefig("Stabilized Return Pressure PL .png")
    plt.show()

def draw_picture():
    global P1_list_stable2, P_list_stable2, QE_list_stable2,\
    QPV_list_stable2, SP_list_stable2,\
    UP_list_stable2,\
    QT_list, QE2_list_stable2,QT_list_stable2,PL_list_stable2,\
    QPV_list_ni
    """Calculate QT_list"""
    QT_list = [QI*QQ + a - b for a, b in zip(QE_list, QE2_list)]

    QPV_list_ni=[-x for x in QPV_list]

    """Extract the stabilized data, I1"""
    P1_list_stable = P1_list[-I1:]
    P_list_stable = P_list[-I1:]
    PL_list_stable2=PL_list[-I1:]+PL_list[-I1:]+PL_list[-I1:]
    QT_list_stable2=QT_list[-I1:]+QT_list[-I1:]+QT_list[-I1:]
    QE_list_stable = QE_list[-I1:]
    QPV_list_stable = QPV_list_ni[-I1:]
    SP_list_stable = SP_list[-I1:]
    UP_list_stable = UP_list[-I1:]
    QE2_list_stable = QE2_list[-I1:]

    """Calculate the time values of the curve after stabilization"""
    T_list_stable = T_list[-I1:]
    T_list_stable1 = [a - T_list_stable[0] for a in T_list_stable]
    T_list_stable2 = [a + T_list_stable1[-1] for a in T_list_stable1]
    T_list_stable3 = [a + T_list_stable2[-1] for a in T_list_stable1]
    T_list_stable0 = T_list_stable1 + T_list_stable2\
                + T_list_stable3
```

```
"""Extract three cycles of the curve after the
 simulation stabilizes"""
P1_list_stable2 = P1_list_stable + P1_list_stable + \
            P1_list_stable
P_list_stable2 = P_list_stable + P_list_stable +\
            P_list_stable
QE_list_stable2 = QE_list_stable + QE_list_stable +\
            QE_list_stable
QPV_list_stable2 = QPV_list_stable + QPV_list_stable +\
            QPV_list_stable
SP_list_stable2 = SP_list_stable + SP_list_stable + \
            SP_list_stable
UP_list_stable2 = UP_list_stable + UP_list_stable + \
            UP_list_stable
QE2_list_stable2 = QE2_list_stable + QE2_list_stable +\
            QE2_list_stable
print("'
       *******     Welcome to the menu     *******
  **** Save the velocity curve, please enter '1'              ****
   **** Save the rear chamber pressure
   curve, please enter '2'       ****
   **** Save the inlet pressure
   curve, please enter '3'           ****
   **** Save the compensation flow
   curve, please enter '4'         ****
   **** Save the curve of the flow required,
   please enter '5'       ****
   **** Save all curves from the entire simulation
   process, enter '6'   ****
   **** Save the curves after the simulation stabilizes,
    please enter '7'****
  **** End the main program,
  please enter 'Q' or other key          ****
       "')
while True:
  flag = input("Please select the figure you want: ")
  if flag == 'Q':
    break
  elif flag == '6':
    PlotDemo_SP_UP_together()
    PlotDemo1(T_list, P1_list)
    PlotDemo1(T_list, QE_list)
    PlotDemo1(T_list, P_list)
    PlotDemo1(T_list, QPV_list_ni)
    PlotDemo1(T_list, PL_list)
    PlotDemo1(T_list, QE2_list)
    PlotDemo1(T_list, QT_list)
    PlotDemo1(T_list, P1_list)
    PlotDemo1(T_list, SP_list)
  elif flag == '1':
    PlotDemo1(T_list, UP_list)
  elif flag == '2':
    PlotDemo1(T_list, P1_list)
  elif flag == '3':
```

```
        PlotDemo1(T_list, P_list)
      elif flag == '4':
        PlotDemo1(T_list, QE_list)
      elif flag == '5':
        PlotDemo1(T_list, QPV_list_ni)
      elif flag == '7':
        PlotDemo_Stable(T_list_stable0, P1_list_stable2)
        PlotDemo_Stable(T_list_stable0, P_list_stable2)
        PlotDemo_Stable(T_list_stable0, QE_list_stable2)
        PlotDemo_Stable(T_list_stable0, QPV_list_stable2)
        PlotDemo_Stable(T_list_stable0, SP_list_stable2)
        PlotDemo_Stable(T_list_stable0, UP_list_stable2)
        PlotDemo_Stable(T_list_stable0, QE2_list_stable2)
        PlotDemo_Stable(T_list_stable0, PL_list_stable2)
        PlotDemo_Stable(T_list_stable0, QT_list_stable2)
      else:
        break

  def initail():
    excel_file = input("Please enter the name and path of "
                "the parameter settings file (.xlsx)")
    parameters_dict = pm.read_excel_parameters(excel_file)

    NN = 680
    NZ = 15
    QI, PA, PLA, K, TT, DTV = pm.parameters70(parameters_dict)
    (PII, TAUP, TAUV, MU, EPSP, EPSV, RHO, ZT1,
    ZT2, ZT3, ZT4, ZT5, ZT6, C)=\
    pm.parameters90(parameters_dict)
    D, H, L, L6, D1, H1, L1, D2, H2, L2 = pm.parameters110\
        (parameters_dict)
    D3, H3, L3, D4, H4, L4, D5, H5, L5 = pm.parameters130 \
          (parameters_dict)
    MP, MV, TST = pm.parameters150(parameters_dict)
    D0, L0, DH, LH, D7, D8 = pm.parameters170(parameters_dict)
    VA, VLA, V20 = pm.parameters190(parameters_dict)
    Sc, Sic, SV, Z0 = pm.parameters210(parameters_dict)
    KL, KC, AA = pm.parameters230(parameters_dict)
    LAMD, QD, PD = pm.parameters250(parameters_dict)
    MM, QQ, Q = pm.parameters270()

    QI, PA, PLA, VA, VLA = pm.coefficient280(QI, PA, PLA, VA,
                          VLA, MM, QQ)
    DTV, V20 = pm.coefficient290(DTV, V20, MM)
    A0, A7, A8 = pm.coefficient300(PII, D0, D7, D8)
    M0, CH, CL = pm.coefficient310(RHO, A0, L0, PA, VA, PLA, VLA)
    V10, VK0, VK = pm.coefficient320(PII, DH, LH, V20)
    KE, K2K = pm.coefficient330(VK0, KC, KL, V20, V10)
    KV = pm.coefficient340(VK, KE)
    KLM, KT = pm.coefficient350(LAMD, RHO, L0, D0, QD, QQ, MM, PD)
    KTR, K10 = pm.coefficient360(A0, KT, KLM)
    B0P, B0V = pm.coefficient370(PII, MU, EPSP, EPSV)
    B1, B2, B3, B4, B5, B6, B7, B8, B9 = pm.coefficient380_420\
     (B0P, D, L, H, L6, D1, L1, H1, D2, L2, H2, TAUP, B0V,
     D3, L3,H3, D4, L4, H4, D5, L5, H5, TAUV, PII)
```

```
 C7, C8 = pm.coefficient430(C, PII, D4, RHO)
 S1V, S2V, S3V, DTZ = pm.coefficient440(SV, Z0, NZ)
 R0P, R0V, R1, R2, R3, R4, R5 = pm.coefficient450_480\
  (PII, EPSP, MU, EPSV, D, H, L6, D1, H1, L1, L, D2, H2,
  L2, D3, H3, L3, D4, H4, L4, D5, H5, L5)
 K1, K2, K3, K4, K5, K6 = pm.coefficient490_500\
  (ZT1, ZT2, ZT3, ZT4, ZT5, ZT6, RHO, A7, A8)
 A1, A2, A3, A5 = pm.coefficient510_520\
  (PII, D, H, D1, H1, D2, H2, D3, H3, D5, H5)
 PH, VH, VL, DT, AK, P2, PL, P1, P, VH1 = pm.coefficient530_540\
  (PA, VA, VLA, TT, PLA)
print("B1:{:<20.7f}B2:{:<20.12f}B3:{:<20.13f}B4:{:<20.13f}B5:"
 "{:<20.13f}".format(B1, B2, B3, B4, B5))
 print("B6:{:<20.10f}B7:{:<20.10f}B8:{:<20.10f}B9:"
 "{:<20.10f}".format(B6, B7, B8, B9))
 print("R1:{:<20.17f}R2:{:<20.16f}R3:{:<20.17f}R4:"
 "{:<20.16f}R5:{:<20.16f}" .format(R1, R2, R3, R4, R5))
 print("K1:{:<20.7f}K2:{:<20.7f}K3:{:<20.7f}K4:"
 {:<20.7f}K5:{:<20.7f}K6:{:<20.7f}"\
 .format(K1, K2, K3, K4, K5, K6))
 (TV, YV, HF, TEE, U0, Y0, T0, QLV, DYV, DYP,
  FV, FSV, FLV, ET, EV, ES, EK, EL, EC, EO, UIM, J,
  flag, T_list, UP_list,UV_list, SP_list,
  P_list, P1_list, P2_list, PL_list,
 QA_list, QPV_list, QE_list, QE2_list, Q11,
  TZ_list, Z1_list, Z2_list, PP1_list, Q11_list,
  Q12_list) = pm.initialize()
 directtoI = False
 FP, FSP, FLP, WP, YP, UP, P3, AP, AV, QL \
 = 0, 0, 0, 0, 0, 0, 0, 0, 0, 0
 WV, UV, F0, FS0, W0, DY0, PPH, KA, QPV, QA, QE, QE2\
 = 0, 0, 0, 0, 0, 0, 0, 0, 0, 0, 0, 0
 Q1, Q2, Q3, Q4, Q5, ZZ, QL1, QL2, U12, DY1, U1, T, Y1= \
 0, 0, 0, 0, 0, 0, 0, 0, 0,\ 0, 0, 0, 0
 S, W1, U22, U2, DY2, Y2, W2, TB, SP = 0, 0, 0, 0, 0, 0, 0, 0, 0
 A11, X1, X2, X3, X4, X5, X6 = 0, 0, 0, 0, 0, 0, 0
 I, Z2, Z1, UVQ, Z, EE, FF, GG, Q12, QP, TZ0, VZ, EZL =\
  0, 0, 0, 0, 0, 0, 0, 0, 0, 0, 0, 0, 0
 TR = 0
 I1 = 0
 TVI = 0
 N = 0
 I2 = 0
 TI = 0
 TTT = 0
 PP1 = 0
 AS = "G"
 TS = 0
 TTI = 0
 YYP = 0
 TVR = 0
 SR = 0
 URM = 0
 SB = 0
```

```
    PMAX = 0
    PMIN = 0
    TVD = 0
    YVD = 0
    print("A1:{:<20.11f}A2:{:<20.11f}A3:{:<20.11f}"
    .format(A1, A2, A3))
    print('state:  {:<10s} {:<10s} {:<10s} {:<10s} {:<10s} {:<10s}'
     '{:<10s}'\.format("PH", "P1", "PL", "UP", "YP", "UV", "YV"))

    return (NN, NZ, QI, PA, PLA, K, TT, DTV, PII, TAUP, TAUV, MU, EPSP,
    EPSV, RHO, ZT1, ZT2, ZT3, ZT4, ZT5, ZT6, C, D, H, L, L6, D1, H1, L1,
    D2, H2, L2, D3, H3, L3, D4, H4, L4, D5, H5, L5, MP, MV, TST, D0, L0,
    DH, LH, D7, D8, VA, VLA, V20, Sc, Sic, SV, Z0, KL, KC, AA, LAMD,
    QD, PD, MM, QQ, Q, A0, A7, A8, M0, CH, CL, V10, VK0, VK, KE,
    K2K, KV, KLM, KT, KTR, K10, B0P, B0V, B1, B2, B3, B4, B5, B6, B7,
    B8, B9, C7, C8, S1V, S2V, S3V, DTZ, R0P, R0V, R1, R2, R3, R4, R5,
    K1, K2, K3, K4, K5, K6, A1, A2, A3, A5, PH, VH, VL, DT, AK, P2,
    PL, P1, P, VH1, TV, YV, HF, TEE, U0, Y0, T0, QLV, DYV, DYP,
    FV, FSV, FLV, ET, EV, ES, EK, EL, EC, EO, UIM, J, flag, T_list,
    UP_list, UV_list, SP_list, P_list, P1_list, P2_list, PL_list,
    QA_list, QPV_list, QE_list, QE2_list, Q11, TZ_list, Z1_list,
    Z2_list, PP1_list, Q11_list, Q12_list, directtoI, FP, FSP, FLP,
    WP, YP, UP, P3, AP, AV, QL, WV, UV, F0, FS0, W0, DY0, PPH, KA, QPV,
     QA, QE, QE2, Q1, Q2, Q3, Q4, Q5, ZZ, QL1, QL2, U12, DY1, U1, T,
     Y1, S, W1, U22, U2, DY2, Y2, W2, TB, SP, A11, X1, X2, X3, X4,
    X5, X6, I, Z2, Z1, UVQ, Z, EE, FF, GG, Q12, QP, TZ0, VZ, EZL, TR,
    I1, TVI, N, I2, TI, TTT, PP1, AS, TS, TTI, YYP, TVR, SR,
    URM, SB, PMAX, PMIN, TVD, YVD)

def piston_moving_3710():
    "'
    Calculation of the force, acceleration, velocity, and
    displacement of the piston.
    "'
    global FP, FSP, FLP, WP, DYP, YP, UP
    FP = A2 * P2 - A1 * P1
    FSP = B1 * UP - np.sign(UP) * (B2 * (P2 - PL) + B3 * (P1 - PL))
    FLP = np.sign(UP) * (B4 * (P2 - PL) + B5 * (P1 - PL))
    WP = (FP - FSP - FLP) / MP # acceleration of the piston
    DYP = UP * DT + (WP * (DT ** 2)) / 2
    YP = YP + DYP #displacement of the piston
    UP = UP + WP * DT #velocity of the piston

def valve_moving_3780():
    "'
    Calculation of the force, acceleration, velocity, and
    displacement of the spool.
    "'
    global FV, FSV, FLV, WV, DYV, YV, UV
    FV = abs(A3 * (P - PL) - A5 * (P - P3))
    FSV = B6 * UV - B7 * (P - PL)
    FLV = B8 * (P - PL) + B9 * (P - P3)
    WV = (FV - FSV - FLV) / MV # acceleration of the spool
    DYV = UV * DT + (WV * (DT ** 2)) / 2
    YV = YV + DYV #displacement of the spool
```

```
    UV = UV + WV * DT #velocity of the spool

def back_oil_moving_3830():
    "'
    Calculation of the acceleration, velocity, and
    displacement of the medium
     in the oil return pipe.
    "'
    global F0, FS0, W0, DY0, Y0, U0, T0
    F0 = A0 * PL
    FS0 = K10 * (U0 ** 2)
    W0 = (F0 - FS0) / M0
    DY0 = U0 * DT + (W0 * (DT ** 2)) / 2
    Y0 = Y0 + DY0
    U0 = U0 + W0 * DT
    T0 = T0 + DT

def high_accumulator_3880():
    "'
    High-pressure flow balance and high-pressure accumulator
    pressure calculation.
    :return:
    "'
    global PPH, KA, QPV, PH, VH, QA, QE, QE2
    PPH = PH
    KA = AK * CH / ((VH ** 0.4) * (PPH ** 2))
    QPV = (AP * DYP + AV * DYV) / DT
   # QPV:Flow rate required for the
    #      movement of the piston and valve spool.
    PH = PPH + (QI - QPV - QL) * DT / (KA + KV)
    VH = (CH / PH) ** AK
    QA = (-KA * (PH - PPH)) / DT
    QE = (-KV * (PH - PPH)) / DT
    QE2 = K2K * QE

def low_accumulator_3930():
    "'
    return oil accumulator pressure calculation
    "'
    global VL, PL
    VL = VL + A0 * DY0 - A11 * DYP - QL * DT
    if VL > VLA:
       VL = VLA
    PL = CL / (VL ** 1.4)

def computation_of_return_pressure_3980():
    "'
    Calculation of the pressure difference during the return
    acceleration phase.
    "'
    global Q1, Q2, Q3, Q4, Q5, P, P1, P2, P3, ZZ
    Q1 = A1 * UP
    Q2 = A2 * UP
    Q3 = QI
    Q4 = Q3 - Q2
```

```
    Q5 = A5 * UV
    P2 = PH + np.sign(Q4) * K4 * (Q4 ** 2) - K2 * (Q2 ** 2)
    P = P2 + K2 * (Q2 ** 2) + K3 * (Q3 ** 2)
    ZZ = abs(S2V - YV)
    P1 = PL + K6 * (Q1 ** 2) + X2 * C8 * ((Q1 / ZZ) ** 2)
    P3 = X1 * PL + X2 * (P2 - K5 * (Q5 ** 2))

def computation_of_impact_pressure_4040():
    '''
    Calculation of the pressure difference during
    the impact acceleration (return braking) phase.
    '''
    global Q1, Q2, Q3, Q4, P, P1, P2, P3, ZZ
    Q1 = A1 * UP
    Q2 = A2 * UP
    Q3 = QI + Q1 - X3 * Q11
    Q4 = QI + Q1 - Q2 - X3 * Q11
    P2 = PH + np.sign(Q4) * K4 * (Q4 ** 2) - np.sign(Q2) *\
    K2 * (Q2 ** 2); P = P2 + np.sign(Q2) * K2 * (Q2 ** 2) + np.sign(Q3)\
    * K3 * (Q3 ** 2); ZZ = abs(S1V - YV)
    P1 = X4 * (P + np.sign(Q1) * K1 * (Q1 ** 2) + X2 \
    * np.sign(UP) * C8 *  ((Q1 / ZZ) ** 2))
    P3 = X1 * PL + X2 * (P2 - K5 * (Q5 ** 2))
    if P1 < 0:
        P1 = 0

def computation_of_flow_4130():
    '''
    Leakage flow rate calculation.
    '''
    global QL1, QL2, QL, QLV
    QL1 = X5 * (R1 * (P1 - PL) + np.sign(UP) * R2 * UP)
    QL2 = R3 * (P2 - PL) + np.sign(UP) * R4 * UP
    QLV = R5 * (P - PL)
    QL = QL1 + QL2 + QLV

def computation_of_energy_loss_4170():
    '''
    Calculation of various energy losses.
    '''
    global ET, EV, ES, EK, EO, EL, EC, ZZ
    ET = ET + P * QI * DT
    EV = EV + X2 * FV * DYV
    ES = ES + FSP * DYP + X2 * FSV * DYV
    EK = EK + abs(FLP * DYP) + X2 * FLV * DYV
    EO = EO + X6 * A1 * PL * abs(DYP)
    EL = EL + ((P1 - PL) * QL1 + (P2 - PL) * QL2 + (P - PL) * QLV) * DT
    ZZ = abs(S3V - YV)
    if ZZ < Z0:
        ZZ = Z0
    EC = EC + (X5 * K1 + C8 / (ZZ ** 2)) * abs((Q1 ** 3) * DT) + \
        (K2 * abs(Q2 ** 3) + K3 * abs(Q3 ** 3) + K4 * abs(Q4 ** 3) + \
         X2 * K5 * abs(Q5 ** 3)) * DT

def computation_of_revision_I_4240():
```

```
    "'
    Calculation of corrections for piston or spool movement alone.
    "'
    global U12, DY1, U1, DT, T, Y1, DYP, DYV, VH, PH
    U12 = U1
    DY1 = Y1 - S
    U1 = np.sign(U1) * (((U12 ** 2) - 2 * abs(W1 * DY1)) ** 0.5)
    DT = (U12 - U1) / W1
    T = T - DT
    Y1 = S
    if W1 == WP:
        DYP = DY1
    else:
        DYV = DY1
    VH = VH - AP * DYP - AV * DYV + (QI - QL - QE) * DT
    PH = CH / (VH ** 1.4)

def computation_of_revision_II_4300():
    "'
    Correction calculation for simultaneous
    movement of the piston and valve spool.
    "'
    global U12, DY1, U1, DT, T, Y1, U22, U2, DY2, Y2, DYP, DYV, VH, PH
    U12 = U1
    DY1 = Y1 - S
    U1 = np.sign(U1) * (((U12 ** 2) - 2 * abs(W1 * DY1)) ** 0.5)
    DT = abs((U12 - U1) / W1)
    T = T - DT
    Y1 = S
    U22 = U2
    U2 = U22 - W2 * DT
    DY2 = ((U22 * U22 - U2 * U2) / W2)/ 2
    Y2 = Y2 - DY2
    if W1 == WP:
        DYP = DY1
        DYV = DY2
    else:
        DYP = DY2
        DYV = DY1
    VH = VH - AP * DYP - AV * DYV + (QI - QL - QE) * DT
    PH = CH / (VH ** 1.4)

def computation_of_revision_III_4370():
    "'
    Correction calculation when the piston
    decelerates to zero velocity.
    "'
    global DT, DYP, TB, YP, UP, SP, VH, PH
    DT = UP / WP
    DYP = WP * DT * DT / 2
    TB = TB - DT
    YP = YP - DYP
    UP = 0
    SP = YP
    VH = VH - AP * DYP - AV * DYV + (QI - QL - QE) * DT
```

```
    PH = CH / (VH ** 1.4)

def positive_mouth_interval_4420():
    ''
    Calculation when the valve spool is in the positive opening range.
    ''
    global I, Z2, Z1, UVQ, Z, UV, DT, EE, FF, GG, \
     P1, Q11, Q12, QP, TZ0, VZ, EZL
    I = 1
    Z2 = Z2 + DTZ
    Z1 = Z0 - Z2
    UVQ = UV
    Z = Z + 1
    UV = (2 * WV * DTZ + UV * UV) ** 0.5
    DT = (UV - UVQ) / WV
    if Z1==0:
        Z1=0.000000002 #Prevent overflow when Z1 = 0.
    EE=(Z1*Z1+I*Z2*Z2)/Z1/Z2/2
    FF = (C8 * Q1 * Q1 + I * Z2 * Z2 * P) / Z1 / Z2 / 2
    GG = P * P - 4 * I * EE * FF * P + 4 * FF * FF
    while(GG<0):
        I = -1
        EE = (Z1 / Z2 + I * Z2 / Z1) / 2
        FF = (C8 * Q1 * Q1 + I * Z2 * Z2 * P) / Z1 / Z2 / 2
        GG = P * P - 4 * I * EE * FF * P + 4 * FF * FF
    P1 = (2 * EE * FF - I * P - I * (GG ** 0.5)) / (EE * EE - I) / 2
    Q11 = C7 * Z1 * (P1 ** 0.5)
    Q12 = np.sign(P1 - P) * C7 * Z2 * (abs(P - P1)** 0.5)
    QP = UP * A1
    TZ0 = TZ0 + DT
    TZ_list.append(TZ0)
    Z1_list.append(Z1)
    Z2_list.append(Z2)
    PP1_list.append(P1)
    Q11_list.append(Q11)
    Q12_list.append(Q12)
    VZ = VZ + Q11 * DT
    EZL = EZL + Q11 * DT * P1

def gen_riversion_list():
    UP_list[-1] = UP
    UV_list[-1] = UV
    P_list[-1] = P / MM
    P1_list[-1] = P1 / MM
    PL_list[-1] = PL / MM
    QA_list[-1] = QA
    QPV_list[-1] = -QPV * QQ
    QE_list[-1] = QE * QQ
    QE2_list[-1] = QE2 * QQ

def gen_list():
    UP_list.append(UP) #save the velocity of the piston
    UV_list.append(UV) #save the velocity of the valve spool
    P_list.append(P / MM) # save the inlet pressure(PMa)
    P1_list.append(P1 / MM) # save the pressure of the rear-chamber
```

```
    PL_list.append(PL / MM) # save the return oil pressure
    QA_list.append(QA * QQ) # save the charging and discharging
    #oil flow rate of the high-pressure accumulator (L/min)
    QPV_list.append(-QPV * QQ) #  save the flow rate
    #required for the movement
    QE_list.append(QE * QQ) # save compensation flow rate Qe
    QE2_list.append(QE2 * QQ) # save compensation flow rate Qe2
    return

def state_A_paramates():
    global AP, AV, A11, X1, X2, X3, X4, X5, X6, TVI, TV, UV, WV, DT, S,\
     ET, EV, ES, EO, EC, EL
    AP = A2
    AV = 0
    A11 = A1
    X1 = 1
    X2 = 0
    X3 = 0
    X4 = 1
    X5 = 0
    X6 = 1
    TVI = TV
    TV = 0
    UV = 0
    WV = 0
    DT = TT
    S = Sc
    ET = 0
    EV = 0
    ES = 0
    EO = 0
    EC = 0
    EL = 0

def state_B_paramates():
    global AP, AV, A11, X1, X2, X3, X4, X5, X6, TV, UV, DT, YV, TB, N, S
    AP = A2
    AV = A3
    A11 = A1
    X1 = 0
    X2 = 1
    X3 = 0
    X4 = 1
    X5 = 0
    X6 = 1
    TV = 0
    UV = 0
    DT = TT
    YV = 0
    TB = 0
    N = 1
    S = S1V

def state_C_paramates():
    global AP, AV, A11, X1, X2, X3, X4, Z1, Z2, TZ0, Z, VZ, EZL, I2
```

```
    AP = A2 - A1
    AV = A3
    A11 = 0
    X1 = 0
    X2 = 1
    X3 = 1
    X4 = 0
    Z1 = 0
    Z2 = 0
    TZ0 = 0
    Z = 0
    VZ = 0
    EZL = 0
    I2 = I1

def state_D_paramates():
    global AP, AV, A11, X1, X2, X3, X4, X5, X6, DT, S
    AP = A2 - A1
    AV = A3
    A11 = 0
    X1 = 0
    X2 = 1
    X3 = 0
    X4 = 1
    X5 = 1
    X6 = 0
    DT = TT
    S = SV

def state_E_paramates():
    global AP, AV, A11, X1, X2, X3, X4, X5, X6, DT, TV, UV, WV
    AP = A2 - A1
    AV = 0
    A11 = 0
    X1 = 1
    X2 = 0
    X3 = 0
    X4 = 1
    X5 = 1
    X6 = 0
    DT = TT
    TV = 0
    UV = 0
    WV = 0

def state_F_paramates():
    global AP, AV, A11, X1, X2, X3, X4, X5, X6, TI, YP, DT, S, N, TTT
    AP = A2 - A1
    AV = 0
    A11 = 0
    X1 = 1
    X2 = 0
    X3 = 0
    X4 = 1
    X5 = 1
```

```
    X6 = 0
    TI = 0
    YP = 0
    DT = TT
    S = Sic - SP
    N = 0
    TTT = TR + TB

def state_G_paramates():
    global AP, AV, A11, X1, X2, X3, X4, X5, X6, TV, UV, YV, DT, S,\
    N, PP1, AS
    AP = A2 - A1
    AV = A5 - A3
    A11 = 0
    X1 = 0
    X2 = 1
    X3 = 0
    X4 = 1
    X5 = 1
    X6 = 0
    TV = 0
    UV = 0
    YV = 0
    DT = TT
    S = -SP
    N = 1
    PP1 = -1
    AS = "G"

def state_I_paramates():
    global AP, AV, A11, X1, X2, X3, X4, X5, X6, WP, UP, YP, TS, DT, TTT
    AP = 0
    AV = A5 - A3
    A11 = 0
    X1 = 0
    X2 = 1
    X3 = 0
    X4 = 1
    X5 = 1
    X6 = 0
    WP = 0
    UP = 0
    YP = 0
    TS = 0
    DT = TST / 5
    TTT = TR + TB + TI

def state_HI_paramates(): #state H'
    global AP, AV, A11, X1, X2, X3, X4, X5, X6, DT, WV, UV, TTI, YYP
    AP = A2
    AV = 0
    A11 = 0
    X1 = 1
    X2 = 0
    X3 = 0
```

```
    X4 = 1
    X5 = 0
    X6 = 1
    DT = TT
    WV = 0
    UV = 0
    TTI = TI
    YYP = YP

def state_II_paramates(): # state I'(1)
    global AP, A11, UP, YP, TS, DT, TTT, WP, WV, AV,\
     X1, X2, X3, X4, X5, X6
    AP = 0
    A11 = 0
    UP = 0
    YP = 0
    TS = 0
    DT = TST / 5
    TTT = TR + TB + TI
    WP = 0
    WV = 0
    AV = 0
    X1 = 1
    X2 = 0
    X3 = 0
    X4 = 1
    X5 = 1
    X6 = 0

def state_II_paramates_2680(): # state I'(2)
    global WP, WV, AV, X1, X2, X3, X4, X5, X6
    WP = 0
    WV = 0
    AV = 0
    X1 = 1
    X2 = 0
    X3 = 0
    X4 = 1
    X5 = 1
    X6 = 0

def state_AI_paramates(): # state A'
    global AP, AV, A11, X1, X2, X3, X4, X5, X6, DT, S, YV
    AP = A2
    AV = A5 - A3
    A11 = A1
    X1 = 0
    X2 = 1
    X3 = 0
    X4 = 1
    X5 = 0
    X6 = 1
    DT = TT
    S = SV
    YV = YV + DYV
```

```
def state_AII_paramates(): # state A"
    global AP, AV, A11, X1, X2, X3, X4, X5, X6, UP, YP, DT, S
    AP = 0
    AV = A5 - A3
    A11 = 0
    X1 = 0
    X2 = 1
    X3 = 0
    X4 = 1
    X5 = 0
    X6 = 1
    UP = 0
    YP = 0
    DT = TT
    S = S2V

def A_state():
    global NN, NZ, QI, PA, PLA, K, TT, DTV, PII, TAUP, TAUV, MU,\
    EPSP, EPSV, RHO, ZT1, ZT2, ZT3, ZT4, ZT5, ZT6, C, D, H, L, L6,\
    D1, H1, L1, D2, H2, L2, D3, H3, L3, D4, H4, L4, D5, H5, L5, MP, MV,\
    TST, D0, L0, DH, LH, D7, D8, VA, VLA, V20, Sc, Sic, SV, Z0, KL, KC,\
    AA, LAMD, QD, PD, MM, QQ, Q, A0, A7, A8, M0, CH, CL, V10, VK0,\
    VK, KE, K2K, KV, KLM, KT, KTR, K10, B0P, B0V, B1, B2, B3, B4, B5,\
    B6, B7, B8, B9, C7, C8, S1V, S2V, S3V, DTZ, R0P, R0V, R1, R2, R3,\
    R4, R5, K1, K2, K3, K4, K5, K6, A1, A2, A3, A5, PH, VH, VL,\
    DT, AK, P2, PL, P1, P, VH1, TV, YV, HF, TEE, U0, Y0, T0, QLV,\
    DYV, DYP, FV, FSV, FLV, ET, EV, ES, EK, EL, EC, EO, UIM, J, flag,\
    T_list, UP_list, UV_list, SP_list, P_list, P1_list, P2_list,\
    PL_list, QA_list, QPV_list, QE_list, QE2_list, Q11, TZ_list,\
    Z1_list, Z2_list, PP1_list, Q11_list, Q12_list, directtoI,\
    FP, FSP, FLP, WP, YP, UP, P3, AP, AV, QL, WV, UV, F0,\
    FS0, W0, DY0, PPH, KA, QPV, QA, QE, QE2, Q1, Q2, Q3, Q4, Q5, ZZ,\
    QL1, QL2, U12, DY1, U1, T, Y1, S, W1, U22, U2, DY2, Y2, W2, TB, \
    SP, A11, X1, X2, X3, X4, X5, X6, I, Z2, Z1, UVQ, Z, EE, FF, GG, \
    Q12, QP, TZ0, VZ, EZL, TR, I1, TVI, N, I2, TI, TTT, PP1, AS, TS,\
    TTI, YYP, TVR, SR, URM, SB, PMAX, PMIN

    print("State A")
    state_A_paramates()
    while True:
        computation_of_return_pressure_3980()
        piston_moving_3710()
        back_oil_moving_3830()
        computation_of_flow_4130()
        high_accumulator_3880()
        low_accumulator_3930()
        computation_of_energy_loss_4170()
        TR = TR + DT #the time of the return acceleration phase.
        I1 += 1
        T_list.append(TR + TEE)
        #TEE is used to record the time at the end of each cycle.
        SP_list.append(YP) #Piston displacement
        gen_list()
        print('A:   {:<10.5f} {:<10.5f} {:<10.5f} {:<10.5f}'
              '{:<10.5f} {:<10.5f} {:<10.5f}'
```

```
        .format(PH / MM,P1 / MM, PL / MM, UP, Q * YP, UV, Q * YV))
        if YP >= S:
          T = TR
          W1 = WP
          U1 = UP
          Y1 = YP
          computation_of_revision_I_4240()
          computation_of_return_pressure_3980()
          TR = T
          UP = U1
          YP = S
          T_list[-1] = TR + TEE
          SP_list[-1] = YP
          gen_riversion_list()
          print('A:   {:<10.5f} {:<10.5f} {:<10.5f} {:<10.5f} '
        '{:<10.5f} {:<10.5f} {:<10.5f}'
        .format(PH / MM, P1 / MM, PL / MM, UP, Q * YP, UV, Q * YV))
          break
    print("-------------------------------")
  def B_state():
    global NN, NZ, QI, PA, PLA, K, TT, DTV, PII, TAUP, TAUV, MU, EPSP,\
    EPSV, RHO, ZT1, ZT2, ZT3, ZT4, ZT5, ZT6, C, D, H, L, L6, D1, H1,\
    L1, D2, H2, L2, D3, H3, L3, D4, H4, L4, D5, H5, L5, MP, MV, TST,\
    D0, L0, DH, LH, D7, D8, VA, VLA, V20, Sc, Sic, SV, Z0, KL, KC, AA,\
    LAMD, QD, PD, MM, QQ, Q, A0, A7, A8, M0, CH, CL, V10, VK0, VK,\
    KE, K2K, KV, KLM, KT, KTR, K10, B0P, B0V, B1, B2, B3, B4, B5,\
    B6, B7, B8, B9, C7, C8, S1V, S2V, S3V, DTZ, R0P, R0V, R1, R2,\
    R3, R4, R5, K1, K2, K3, K4, K5, K6, A1, A2, A3, A5, PH, VH, VL,\
    DT, AK, P2, PL, P1, P, VH1, TV, YV, HF, TEE, U0, Y0, T0, QLV, DYV, \
    DYP, FV, FSV, FLV, ET, EV, ES, EK, EL, EC, EO, UIM, J, flag,\
    T_list, UP_list, UV_list, SP_list, P_list, P1_list, P2_list,\
    PL_list, QA_list, QPV_list, QE_list, QE2_list, Q11, TZ_list,\
    Z1_list, Z2_list, PP1_list, Q11_list, Q12_list, directtoI, FP,\
    FSP, FLP, WP, YP, UP, P3, AP, AV, QL, WV, UV, F0, FS0, W0, DY0,\
    PPH, KA, QPV, QA, QE, QE2, Q1, Q2, Q3, Q4, Q5, ZZ, QL1, QL2,\
    U12, DY1, U1, T, Y1, S, W1, U22, U2, DY2, Y2, W2, TB, SP, A11,\
    X1, X2, X3, X4, X5, X6, I, Z2, Z1, UVQ, Z, EE, FF, GG, Q12, QP, TZ0,\
    VZ, EZL, TR, I1, TVI, N, I2, TI, TTT, PP1, AS, TS, TTI, YYP,\
    TVR, SR, URM, SB, PMAX, PMIN

    print("State B")
    state_B_paramates()
    while True:
      computation_of_return_pressure_3980()
      piston_moving_3710()
      valve_moving_3780()
      back_oil_moving_3830()
      computation_of_flow_4130()
      high_accumulator_3880()
      low_accumulator_3930()
      computation_of_energy_loss_4170()
      if WP < 0: # Enter the return deceleration phase.
        TB += DT #the time of the return deceleration phase.
        TV += DT #Calculate the return reversal time.
```

```
            I1 += 1
            T_list.append(TR + TB + TEE)
            SP_list.append(YP)
            gen_list()
        else:
            TR += DT
            TV += DT
            SR = YP
            URM = UP
            I1 += 1
            T_list.append(TR + TEE)
            SP_list.append(YP)
            gen_list()
        print('B:  {:<10.5f} {:<10.5f} {:<10.5f} {:<10.5f} {:<10.5f}'
        ' {:<10.5f} {:<10.5f}'.format(PH / MM, P1 / MM, PL /
        MM, UP, Q * YP, UV, Q * YV))
        if YV >= S:
            T = TB
            W1 = WV
            U1 = UV
            Y1 = YV
            W2 = WP
            U2 = UP
            Y2 = YP
            computation_of_revision_II_4300()
            TB = T
            UV = U1
            YV = S
            UP = U2
            YP = Y2
            computation_of_return_pressure_3980()
            T_list[-1] = TR + TB + TEE
            SP_list[-1] = YP
            gen_riversion_list()
            print('B:  {:<10.5f} {:<10.5f} {:<10.5f}'
            ' {:<10.5f} {:<10.5f} {:<10.5f} {:<10.5f} '\
            .format(PH / MM, P1 / MM, PL / MM, UP,
            Q * YP, UV, Q * YV))
            print("Tr={:<8s} ms, Tb={:<8s} ms, Sr={:<8s} mm,"
            "Urm={:<8s} m/s".format(format_value(Q * TR),
            format_value(Q * TB), format_value(Q * SR),
            format_value(URM)))
            break
        else:
            continue
    print("-----------------------------")
def C_state():
    global NN, NZ, QI, PA, PLA, K, TT, DTV, PII, TAUP, TAUV, MU, EPSP,\
    EPSV, RHO, ZT1, ZT2, ZT3, ZT4, ZT5, ZT6, C, D, H, L, L6, D1, H1,\
    L1, D2, H2, L2, D3, H3, L3, D4, H4, L4, D5, H5, L5, MP, MV, TST,\
    D0, L0, DH, LH, D7, D8, VA, VLA, V20, Sc, Sic, SV, Z0, KL, KC, \
    AA, LAMD, QD, PD, MM, QQ, Q, A0, A7, A8, M0, CH, CL, V10, VK0, VK,\
    KE, K2K, KV, KLM, KT, KTR, K10, B0P, B0V, B1, B2, B3, B4, B5, B6,\
    B7, B8, B9, C7, C8, S1V, S2V, S3V, DTZ, R0P, R0V, R1, R2, R3, R4,\
```

```
    R5, K1, K2, K3, K4, K5, K6, A1, A2, A3, A5, PH, VH, VL, DT, AK,\
    P2, PL, P1, P, VH1, TV, YV, HF, TEE, U0, Y0, T0, QLV, DYV, DYP,\
    FV, FSV, FLV, ET, EV, ES, EK, EL, EC, EO, UIM, J, flag,\
    T_list, UP_list, UV_list, SP_list, P_list, P1_list, P2_list,\
    PL_list, QA_list, QPV_list, QE_list, QE2_list, Q11, TZ_list,\
    Z1_list, Z2_list, PP1_list, Q11_list, Q12_list, directtoI, FP,\
    FSP, FLP, WP, YP,  UP, P3, AP, AV, QL, WV, UV, F0, FS0, W0, DY0,\
    PPH, KA, QPV, QA, QE, QE2, Q1, Q2, Q3, Q4, Q5, ZZ, QL1, QL2,\
    U12, DY1, U1, T, Y1, S, W1, U22, U2, DY2, Y2, W2, TB, SP, A11,\
    X1, X2, X3, X4, X5, X6, I, Z2, Z1, UVQ, Z, EE, FF, GG, Q12, QP, TZ0,\
    VZ, EZL, TR, I1, TVI, N, I2, TI, TTT, PP1, AS, TS, TTI, YYP,\
    TVR, SR, URM, SB, PMAX, PMIN

    print("State C:")
    state_C_paramates()
    computation_of_return_pressure_3980()
    while True:
      if Z != 0:
        computation_of_impact_pressure_4040()
      positive_mouth_interval_4420()
      piston_moving_3710()
      valve_moving_3780()
      YV = S1V + DTZ * Z
      back_oil_moving_3830()
      computation_of_flow_4130()
      high_accumulator_3880()
      low_accumulator_3930()
      TB += DT
      TV += DT
      I1 += 1
      T_list.append(TR + TB + TEE)
      SP_list.append(YP)
      gen_list()
     print('C:  {:<10.5f} {:<10.5f} {:<10.5f} {:<10.5f} {:<10.5f}'
      '{:<10.5f} {:<10.5f}'.format(PH / MM, P1 / MM, PL / MM, UP,
      Q * YP, UV, Q * YV))
      if Z >= NZ:
        break
    print("-----------------------------")
def D_state():
    global NN, NZ, QI, PA, PLA, K, TT, DTV, PII, TAUP, TAUV, MU, EPSP,\
    EPSV, RHO, ZT1, ZT2, ZT3, ZT4, ZT5, ZT6, C, D, H, L, L6, D1, H1,\
    L1, D2, H2, L2, D3, H3, L3, D4, H4, L4, D5, H5, L5, MP, MV, TST,\
     D0, L0, DH, LH, D7, D8, VA, VLA, V20, Sc, Sic, SV, Z0, KL, KC,\
    AA, LAMD, QD, PD, MM, QQ, Q, A0, A7, A8, M0, CH, CL, V10, VK0, VK, \
    KE, K2K, KV, KLM, KT, KTR, K10, B0P, B0V, B1, B2, B3, B4, B5, B6,\
    B7, B8, B9, C7, C8, S1V, S2V, S3V, DTZ, R0P, R0V, R1, R2, R3, R4, \
    R5, K1, K2, K3, K4, K5, K6, A1, A2, A3, A5, PH, VH, VL, DT, AK, P2,\
    PL, P1, P, VH1, TV, YV, HF, TEE, U0, Y0, T0, QLV, DYV, DYP, FV, \
    FSV, FLV, ET, EV, ES, EK, EL, EC, EO, UIM, J, flag, T_list, \
    UP_list, UV_list, SP_list, P_list, P1_list, P2_list, PL_list,\
    QA_list, QPV_list, QE_list, QE2_list, Q11, TZ_list, Z1_list,\
    Z2_list, PP1_list, Q11_list, Q12_list, directtoI, FP, FSP,\
    FLP, WP, YP, UP, P3, AP, AV, QL, WV, UV, F0, FS0, W0, DY0, PPH, KA,\
```

```
QPV, QA, QE, QE2, Q1, Q2, Q3, Q4, Q5, ZZ, QL1, QL2, U12, DY1, U1,\
T, Y1, S W1, U22, U2, DY2, Y2, W2, TB, SP, A11, X1, X2, X3, \
X4, X5, X6, I, Z2, Z1, UVQ, Z, EE, FF, GG, Q12, QP, TZ0,\
VZ, EZL, TR, I1, TVI, N, I2, TI, TTT, PP1, AS, TS, TTI,\
YYP, TVR, SR, URM, SB, PMAX, PMIN

print("State D:")
state_D_paramates()
while True:
  piston_moving_3710()
  computation_of_impact_pressure_4040()
  valve_moving_3780()
  back_oil_moving_3830()
  computation_of_flow_4130()
  high_accumulator_3880()
  low_accumulator_3930()
  computation_of_energy_loss_4170()
  TB += DT
  TV += DT
  I1 += 1
  T_list.append(TR + TB + TEE)
  SP_list.append(YP)
  gen_list()
  print('D:   {:<10.5f} {:<10.5f} {:<10.5f} {:<10.5f} '
  '{:<10.5f} {:<10.5f} {:<10.5f}'.format(PH / MM, P1 / MM, PL
  / MM, UP, Q * YP, UV, Q * YV))
  if YV >= S:
    T = TB
    W1 = WV
    U1 = UV
    Y1 = YV
    W2 = WP
    U2 = UP
    Y2 = YP
    computation_of_revision_II_4300()
    computation_of_impact_pressure_4040()
    TB = T
    UV = U1
    YV = S
    UP = U2
    YP = Y2
    T_list[-1] = TR + TB + TEE
    SP_list[-1] = YP
    gen_riversion_list()
    print('D:   {:<10.5f} {:<10.5f} {:<10.5f} {:<10.5f}'
    '{:<10.5f} {:<10.5f} {:<10.5f}'.format(PH / MM, P1 / MM, PL
    / MM, UP, Q * YP, UV, Q * YV))
    print('Tr={:<8s} ms  Tv={:<8s} ms  Tb={:<8s} ms'.format(
      format_value(TR * Q),
      format_value(TV * Q),
      format_value(TB * Q)
    ))
    TVR = TV #Record the valve spool return reversal time.
    TV = 0
```

```
        break
    print("------------------------------")
  def E_state():
    global NN, NZ, QI, PA, PLA, K, TT, DTV, PII, TAUP, TAUV, MU, \
    EPSP, EPSV, RHO, ZT1, ZT2, ZT3, ZT4, ZT5, ZT6, C, D, H, L, L6,\
    D1, H1, L1, D2, H2, L2, D3, H3, L3, D4, H4, L4, D5, H5, L5, MP, MV,\
    TST, D0, L0, DH, LH, D7, D8, VA, VLA, V20, Sc, Sic, SV, Z0, KL,\
    KC, AA, LAMD, QD, PD, MM, QQ, Q, A0, A7, A8, M0, CH, CL, V10, VK0,\
    VK, KE, K2K, KV, KLM, KT, KTR, K10, B0P, B0V, B1, B2, B3, B4, B5,\
    B6, B7, B8, B9, C7, C8, S1V, S2V, S3V, DTZ, R0P, R0V, R1, R2, R3,\
    R4, R5, K1, K2, K3, K4, K5, K6, A1, A2, A3, A5, PH, VH, VL, DT,\
    AK, P2, PL, P1, P, VH1, TV, YV, HF, TEE, U0, Y0, T0, QLV, DYV, \
    DYP, FV, FSV, FLV, ET, EV, ES, EK, EL, EC, EO, UIM, J, flag, \
    T_list, UP_list, UV_list, SP_list, P_list, P1_list, P2_list,\
    PL_list, QA_list, QPV_list, QE_list, QE2_list, Q11, TZ_list,\
    Z1_list, Z2_list, PP1_list, Q11_list, Q12_list, directtoI, FP,\
    FSP, FLP, WP, YP, UP, P3, AP, AV, QL, WV, UV, F0, FS0, W0, DY0,\
    PPH, KA, QPV, QA, QE, QE2, Q1, Q2, Q3, Q4, Q5, ZZ, QL1, QL2, U12,\
    DY1, U1, T, Y1, S, W1, U22, U2, DY2, Y2, W2, TB, SP, A11, X1, X2,\
     X3, X4, X5, X6, I, Z2, Z1, UVQ, Z, EE, FF, GG, Q12, QP, TZ0,\
    VZ, EZL, TR, I1, TVI, N, I2, TI, TTT, PP1, AS, TS, TTI, YYP,\
    TVR, SR, URM, SB, PMAX, PMIN

    print("State E:")
    state_E_paramates()
    while True:
      computation_of_impact_pressure_4040()
      piston_moving_3710()
      back_oil_moving_3830()
      computation_of_flow_4130()
      high_accumulator_3880()
      low_accumulator_3930()
      computation_of_energy_loss_4170()
      TB = TB + DT
      I1 = I1 + 1
      T_list.append(TR + TB + TEE)
      SP_list.append(YP)
      gen_list()
      print('E:   {:<10.5f} {:<10.5f} {:<10.5f} {:<10.5f} '
     ' {:<10.5f} {:<10.5f} {:<10.5f}'.format(PH / MM, P1 / MM, PL /
      MM, UP, Q * YP,UV, Q * YV))
      if UP <= 0:
        computation_of_revision_III_4370()
        computation_of_impact_pressure_4040()
        T_list[-1] = TR + TB + TEE
        SP_list[-1] = YP
        gen_riversion_list()
      print('E:  {:<10.5f} {:<10.5f} {:<10.5f} {:<10.5f} {:<10.5f} '
          '{:<10.5f} {:<10.5f}'.format(PH / MM, P1 /
         MM, PL / MM, UP, Q * YP, UV, Q * YV))
        SB = SP - SR #Record the displacement during
       # the return deceleration phase.
        print('Tb={:<8s} ms; Sp={:<8s} mm; Sb={:<8s} mm;
         Sr={:<8s} mm'.format(
```

```
            format_value(Q * TB),
            format_value(Q * SP),
            format_value(Q * SB),
            format_value(Q * SR)
        ))
        break
    print("------------------------------")
def F_state():
    global NN, NZ, QI, PA, PLA, K, TT, DTV, PII, TAUP, TAUV, MU, EPSP,\
    EPSV, RHO, ZT1, ZT2, ZT3, ZT4, ZT5, ZT6, C, D, H, L, L6, D1,\
    H1, L1, D2, H2, L2, D3, H3, L3, D4, H4, L4, D5, H5, L5, MP, MV,\
    TST, D0, L0, DH, LH, D7, D8, VA, VLA, V20, Sc, Sic, SV, Z0, KL,\
    KC, AA, LAMD, QD, PD, MM, QQ, Q, A0, A7, A8, M0, CH, CL, V10, VK0,\
    VK, KE, K2K, KV, KLM, KT, KTR, K10, B0P, B0V, B1, B2,\
    B3, B4, B5, B6, B7, B8, B9, C7, C8, S1V, S2V, S3V, DTZ, R0P, R0V,\
    R1, R2, R3, R4, R5, K1, K2, K3, K4, K5, K6, A1, A2, A3, A5, PH, \
    VH, VL, DT, AK, P2, PL, P1, P, VH1, TV, YV, HF, TEE, U0, Y0,\
    T0, QLV,\
    DYV, DYP, FV, FSV, FLV, ET, EV, ES, EK, EL, EC, EO, UIM, J, flag, \
    T_list, UP_list, UV_list, SP_list, P_list, P1_list, P2_list,\
    PL_list, QA_list, QPV_list, QE_list, QE2_list, Q11, TZ_list,\
    Z1_list, Z2_list, PP1_list, Q11_list, Q12_list, directtoI, FP,\
    FSP, FLP, WP, YP, UP, P3, AP, AV, QL, WV, UV, F0, FS0, W0, DY0,\
    PPH, KA, QPV, QA, QE, QE2, Q1, Q2, Q3, Q4, Q5, ZZ, QL1, QL2, \
    U12, DY1, U1, T, Y1, S, W1, U22, U2, DY2, Y2, W2, TB, SP, \
    A11, X1, X2, X3, X4, X5, X6, I, Z2, Z1, UVQ, Z, EE, FF,\
    GG, Q12, QP, TZ0, VZ, EZL, TR, I1, TVI, N, I2, TI, TTT,\
    PP1, AS, TS, TTI, YYP, TVR, SR, URM, SB, PMAX, PMIN

    print("State F:")
    state_F_paramates()
    while True:
        computation_of_impact_pressure_4040()
        piston_moving_3710()
        back_oil_moving_3830()
        computation_of_flow_4130()
        high_accumulator_3880()
        low_accumulator_3930()
        computation_of_energy_loss_4170()
        if N <= 0:
            if QA < 0:
                PMAX = PH
                N = 1
        TI = TI + DT
        I1 += 1
        T_list.append(TTT + TI + TEE)
        SP_list.append(SP + YP)
        gen_list()
        print('F: {:<10.5f} {:<10.5f} {:<10.5f} {:<10.5f} {:<10.5f}'
        ' {:<10.5f} {:<10.5f}'.format(PH / MM, P1 / MM, PL / MM, UP,
         Q * YP, UV, Q * YV))
        if YP <= S:
            T = TI
            W1 = WP
```

```
            U1 = UP
            Y1 = YP
            computation_of_revision_I_4240()
            computation_of_impact_pressure_4040()
            TI = T
            UP = U1
            YP = S
            T_list[-1] = TTT + TI + TEE
            SP_list[-1] = SP + YP
            gen_riversion_list()
          print('F:{:<10.5f} {:<10.5f} {:<10.5f} {:<10.5f} {:<10.5f}'
            '{:<10.5f} {:<10.5f}'.format(PH / MM, P1 / MM, PL / MM,
             UP, Q * YP, UV, Q * YV))
            break
    print("------------------------------")
def G_state():
    global NN, NZ, QI, PA, PLA, K, TT, DTV, PII, TAUP, TAUV, MU, EPSP,\
    EPSV, RHO, ZT1, ZT2, ZT3, ZT4, ZT5, ZT6, C, D, H, L, L6, D1, H1,\
    L1, D2, H2, L2, D3, H3, L3, D4, H4, L4, D5, H5, L5, MP, MV, TST,\
    D0, L0, DH, LH, D7, D8, VA, VLA, V20, Sc, Sic, SV, Z0, KL, KC,\
    AA, LAMD, QD, PD, MM, QQ, Q, A0, A7, A8, M0, CH, CL, V10,\
    VK0, VK, KE, K2K, KV, KLM, KT, KTR, K10, B0P, B0V, B1, B2,\
    B3, B4, B5, B6, B7, B8, B9, C7, C8, S1V, S2V, S3V,\
    DTZ, R0P, R0V,  R1, R2, R3, R4, R5, K1, K2, K3, K4, K5, K6,\
    A1, A2, A3, A5, PH,  VH, VL, DT, AK, P2, PL, P1, P, VH1, TV,\
    YV, HF, TEE, U0, Y0, T0, QLV, DYV, DYP, FV, FSV, FLV, ET, EV,\
    ES, EK, EL, EC, EO, UIM, J, flag, T_list, UP_list, UV_list,\
    SP_list, P_list, P1_list, P2_list, PL_list, QA_list, QPV_list,\
    QE_list, QE2_list, Q11, TZ_list, Z1_list, Z2_list,\
    PP1_list, Q11_list, Q12_list, directtoI, FP, FSP, FLP, WP, YP,\
    UP, P3, AP, AV, QL, WV, UV, F0, FS0, W0, DY0, PPH, KA, QPV, QA,\
    QE, QE2, Q1, Q2, Q3, Q4, Q5, ZZ, QL1, QL2, U12,\
    DY1, U1, T, Y1, S, W1, U22, U2, DY2, Y2, W2, TB, SP, \
    A11, X1, X2, X3, X4, X5, X6, I, Z2, Z1, UVQ, Z, EE, FF, GG, Q12, QP,\
    TZ0, VZ, EZL, TR, I1, TVI, N, I2, TI, TTT, PP1, AS, TS, TTI, YYP,\
    TVR, SR, URM, SB, PMAX, PMIN

    print("State G:")
    state_G_paramates()
    while True:
        if N > 0:
            computation_of_impact_pressure_4040()
        else:
            computation_of_return_pressure_3980()

        piston_moving_3710()
        valve_moving_3780()
        back_oil_moving_3830()
        computation_of_flow_4130()
        high_accumulator_3880()
        low_accumulator_3930()
        computation_of_energy_loss_4170()
        TI = TI + DT # Calculate the time of the impact stroke phase.
        TV = TV + DT # Calculate the impact reversal time.
        I1 = I1 + 1
```

```
T_list.append(TTT + TI + TEE)
SP_list.append(SP + YP)
gen_list()
print('{}: {:<10.5f} {:<10.5f} {:<10.5f} {:<10.5f}'
'{:<10.5f} {:<10.5f} {:<10.5f}'.format(AS, PH / MM, P1 /
MM, PL / MM, UP, Q * YP, UV, Q * YV))
if WP > 0:
   N = 0
   AP = A2
   X1 = 1
   X2 = 0
   AS = "H"
if YV > SV:
   T = TI
   W1 = WV
   U1 = UV
   Y1 = YV
   S = SV
   W2 = WP
   U2 = UP
   Y2 = YP
   computation_of_revision_II_4300()
   computation_of_impact_pressure_4040()
   TI = T
   TVI = TV - DT #Record the valve spool impact reversal time.
   UV = U1
   YV = S
   UP = U2
   YP = Y2
   S = -SP
   T_list[-1] = TTT + TI + TEE
   SP_list[-1] = SP + YP
   gen_riversion_list()
   print('H:   {:<10.5f} {:<10.5f} {:<10.5f}'
   '{:<10.5f} {:<10.5f} {:<10.5f} {:<10.5f}'
   .format(PH / MM, P1 / MM, PL / MM, UP, Q * YP, UV, Q * YV))
   print("Valve Stop. Tvi={}".format(TVI * Q))
   print("-----------------------------")
   directtoI = False
   break
if YP <= S: #Piston completes the impact,
#and valve spool moves. I state
   T = TI
   W1 = WP
   U1 = UP
   Y1 = YP
   W2 = WV
   U2 = UV
   Y2 = YV
   computation_of_revision_II_4300()
   computation_of_impact_pressure_4040()
   if YV < S3V:
      P1 = P1 + AA * RHO * abs(UP)
      PP1 = 1
```

```
            TI = T
            TV -= DT
            UP = U1
            YP = S
            UV = U2
            YV = Y2
            UIM = -UP
            PMIN = P
            T_list[-1] = TTT + TI + TEE
            SP_list[-1] = SP + YP
            gen_riversion_list()
            print('{}: {:<10.5f} {:<10.5f} {:<10.5f} {:<10.5f}'
            '{:<10.5f} {:<10.5f} {:<10.5f}'.format(AS, PH / MM, P1 /
            MM, PL / MM, UP, Q * YP, UV, Q * YV))
            directtoI = True
            break

def HI_state(): # H' state. Spool stops moving,
    global NN, NZ, QI, PA, PLA, K, TT, DTV, PII, TAUP, TAUV, MU, EPSP,\
    EPSV, RHO, ZT1, ZT2, ZT3, ZT4, ZT5, ZT6, C, D, H, L, L6,\
    D1, H1, L1, D2, H2, L2, D3, H3, L3, D4, H4, L4, D5, H5, L5, MP,\
    MV, TST, D0, L0, DH, LH, D7, D8, VA, VLA, V20, Sc, Sic, SV, \
    Z0, KL, KC, AA, LAMD, QD, PD, MM, QQ, Q, A0, A7, A8, M0, \
    CH, CL, V10, VK0, VK, KE, K2K, KV,  KLM, KT, KTR, K10, \
    B0P, B0V, B1, B2, B3, B4, B5, B6, B7,  B8, B9, C7, C8, \
    S1V, S2V, S3V, DTZ, R0P, R0V, R1, R2, R3, R4, R5, K1, K2, K3, K4,\
    K5, K6, A1, A2, A3, A5, PH, VH, VL, DT, AK, P2, PL, P1, P,\
    VH1, TV, YV, HF, TEE, U0, Y0, T0, QLV,  DYV, DYP, FV, FSV,\
    FLV, ET, EV, ES, EK, EL, EC, EO, UIM, J, flag, T_list, UP_list,\
    UV_list, SP_list, P_list, P1_list, P2_list, PL_list, QA_list, \
    QPV_list, QE_list, QE2_list, Q11, TZ_list, Z1_list, Z2_list,\
     PP1_list, Q11_list, Q12_list, directtoI, FP, FSP, FLP, WP,\
     YP, UP, P3, AP, AV, QL, WV, UV, F0, FS0, W0, DY0, PPH, KA,\
     QPV, QA, QE, QE2, Q1, Q2, Q3, Q4, Q5, ZZ, QL1, QL2, U12, DY1,\
     U1, T, Y1, S, W1, U22, U2, DY2, Y2, W2, TB, SP, A11, X1, X2,\
     X3, X4, X5, X6, I, Z2, Z1, UVQ, Z, EE, FF, GG, Q12, QP,\
     TZ0, VZ,  EZL, TR, I1, TVI, N, I2, TI, TTT, PP1, AS, TS, \
     TTI, YYP, TVR, SR, URM, SB, PMAX, PMIN

    print("State H':")
    state_HI_paramates()
    while True:
        computation_of_return_pressure_3980()
        piston_moving_3710()
        back_oil_moving_3830()
        computation_of_flow_4130()
        high_accumulator_3880()
        low_accumulator_3930()
        computation_of_energy_loss_4170()
        TI = TI + DT
        I1 = I1 + 1
        T_list.append(TTT + TI + TEE)
        SP_list.append(SP + YP)
        gen_list()
        print("H':   {:<10.5f} {:<10.5f} {:<10.5f}"
```

```
        "{:<10.5f} {:<10.5f} {:<10.5f} {:<10.5f}".format(PH / MM, P1 /
        MM, PL / MM, UP, Q * YP, UV, Q * YV))
        if YP <= S: # Piston completes impact
          T = TI
          W1 = WP
          U1 = UP
          Y1 = YP
          computation_of_revision_I_4240()
          computation_of_return_pressure_3980()
          TI = T
          UP = U1
          YP = S
          UIM = -UP
          PMIN = P
          YV = SV
          T_list[-1] = (TTT + TI + TEE)
          SP_list[-1] = (SP + YP)
          gen_riversion_list()
          print("H':   {:<10.5f} {:<10.5f} {:<10.5f}"
          " {:<10.5f} {:<10.5f} {:<10.5f} {:<10.5f}"
          .format(PH / MM, P1 / MM, PL / MM, UP,
          Q * YP, UV, Q * YV))
          print("Valve Lead, Time:{:<8s} ms, Stroke:{:<8s} mm"
          .format(format_value(Q * (TI - TTI)),
          format_value(Q * (YP - YYP))
          ))
          break
    print("-----------------------------")
  def II_state(): #I' state: transition from state H' to state I'
    global NN, NZ, QI, PA, PLA, K, TT, DTV, PII, TAUP, TAUV, MU, EPSP,\
    EPSV, RHO, ZT1, ZT2, ZT3, ZT4, ZT5, ZT6, C, D, H, L, L6, D1, H1,\
    L1, D2, H2, L2, D3, H3, L3, D4, H4, L4, D5, H5, L5, MP, MV, TST,\
    D0, L0, DH, LH, D7, D8, VA, VLA, V20, Sc, Sic, SV, Z0, KL, KC, \
    AA, LAMD, QD, PD, MM, QQ, Q, A0, A7, A8, M0, CH, CL, V10, VK0,\
    VK, KE, K2K, KV, KLM, KT, KTR, K10, B0P, B0V, B1, B2, B3, B4, B5,\
    B6, B7, B8, B9, C7, C8, S1V, S2V, S3V, DTZ, R0P, R0V, R1, \
    R2, R3, R4, R5, K1, K2, K3, K4, K5, K6, A1, A2, A3, A5, PH, VH, VL,\
    DT, AK, P2, PL, P1, P, VH1, TV, YV, HF, TEE, U0, Y0, T0, QLV,\
    DYV, DYP, FV, FSV, FLV, ET, EV, ES, EK, EL, EC, EO, UIM, J, flag,\
    T_list, UP_list, UV_list, SP_list, P_list, P1_list, P2_list, \
    PL_list, QA_list, QPV_list, QE_list, QE2_list, Q11, TZ_list,\
    Z1_list, Z2_list, PP1_list, Q11_list, Q12_list, directtoI,\
    FP, FSP, FLP, WP, YP, UP, P3, AP, AV, QL,\ WV, UV, F0, FS0, W0,\
    DY0, PPH, KA, QPV, QA, QE, QE2, Q1, Q2, Q3, Q4, Q5, ZZ, QL1, QL2,\
    U12, DY1, U1, T, Y1, S, W1, U22, U2, DY2, Y2, W2, TB, SP, A11,\
    X1, X2, X3, X4, X5, X6, I, Z2, Z1, UVQ, Z, EE, FF, GG, Q12, QP, TZ0,\
    VZ, EZL, TR, I1, TVI, N, I2, TI, TTT, PP1, AS, TS, TTI, YYP, TVR,\
    SR, URM, SB, PMAX, PMIN

    print("State I':")
    state_II_paramates()
    while True:
      computation_of_return_pressure_3980()
      back_oil_moving_3830()
```

```
        computation_of_flow_4130()
        high_accumulator_3880()
        low_accumulator_3930()
        ET = ET + P * QI * DT
        EL = EL + ((P1 - PL) * QL1 + (P2 - PL) * QL2 + (P - PL) * QLV) * DT
        TS = TS + DT
        I1 = I1 + 1
        T_list.append(TTT + TS + TEE)
        SP_list.append(YP)
        gen_list()
        print("I':   {:<10.5f} {:<10.5f} {:<10.5f}"
        "{:<10.5f} {:<10.5f} {:<10.5f} {:<10.5f}"
        .format(PH / MM, P1 / MM, PL / MM, UP,
        Q * YP, UV, Q * YV))
        if TS >= TST:
            break
    TTT = TR + TB + TI + TST #Record the time of one cycle.
    print("Ti={:<8s} ms".format(format_value(Q * TI)))
    # display the time of the impact stroke
    print("-----------------------------") # End of one cycle.
    def II_state_2680():#Transition from state I to state I'
    global NN, NZ, QI, PA, PLA, K, TT, DTV, PII, TAUP, TAUV, MU,\
    EPSP, EPSV, RHO, ZT1, ZT2, ZT3, ZT4, ZT5, ZT6, C, D, H, L,\
    L6, D1, H1, L1, D2, H2, L2, D3, H3, L3, D4, H4, L4, D5, H5,\
    L5, MP, MV, TST, D0, L0, DH, LH, D7, D8, VA, VLA, V20, Sc, Sic,\
    SV, Z0, KL, KC, AA, LAMD, QD, PD, MM, QQ, Q, A0, A7,\
    A8, M0, CH, CL, V10, VK0, VK, KE, K2K, KV, KLM, KT, KTR, K10, \
    B0P, B0V, B1, B2, B3, B4, B5, B6, B7, B8, B9, C7, C8, S1V, S2V,\
    S3V, DTZ,  R0P, R0V, R1, R2, R3, R4, R5, K1, K2, K3, K4, K5, K6,\
    A1, A2, A3, A5, PH, VH, VL, DT, AK, P2, PL, P1, P, VH1, TV, YV, \
    HF, TEE, U0, Y0, T0, QLV, DYV, DYP, FV, FSV, FLV, ET, EV, ES, EK,\
    EL, EC, EO, UIM, J, flag, T_list, UP_list, UV_list,\
    SP_list, P_list, P1_list, P2_list, PL_list, QA_list, QPV_list, \
    QE_list, QE2_list, Q11, TZ_list, Z1_list, Z2_list, PP1_list, \
    Q11_list, Q12_list, directtoI, FP, FSP, FLP, WP, YP, UP, P3,\
    AP, AV, QL, WV, UV, F0, FS0, W0, DY0, PPH, KA, QPV, QA, QE, QE2,\
    Q1, Q2, Q3, Q4, Q5, ZZ, QL1, QL2, U12, DY1, U1, T, Y1, S, W1,\
    U22, U2, DY2, Y2, W2, TB, SP, A11, X1, X2, X3, X4, X5, X6, I, Z2,\
    Z1, UVQ, Z, EE, FF, GG, Q12, QP, TZ0, VZ, EZL, TR, I1, TVI, N, I2,\
    TI, TTT, PP1, AS, TS, TTI, YYP, TVR, SR, URM, SB, PMAX, PMIN

    state_II_paramates_2680()
    while True:
        computation_of_return_pressure_3980()
        back_oil_moving_3830()
        computation_of_flow_4130()
        high_accumulator_3880()
        low_accumulator_3930()
        ET = ET + P * QI * DT
        EL = EL + ((P1 - PL) * QL1 + (P2 - PL) * QL2 + (P - PL) * QLV) * DT
        TS = TS + DT
        I1 = I1 + 1
        T_list.append(TTT + TS + TEE)
        SP_list.append(YP)
```

```
            gen_list()
            print("I':   {:<10.5f} {:<10.5f} {:<10.5f} {:<10.5f}"
           " {:<10.5f} {:<10.5f} {:<10.5f}".format(PH / MM, P1 /
            MM, PL / MM, UP, Q * YP, UV, Q * YV))
            if TS >= TST:
                break
        TTT = TR + TB + TI + TST
        print("Ti={:<8s} ms".format(format_value(Q * TI)))
        # display the time of the impact stroke
        print("------------------------------")# End of one cycle.
    def I_state():
        global NN, NZ, QI, PA, PLA, K, TT, DTV, PII, TAUP, TAUV,\
        MU, EPSP, EPSV, RHO, ZT1, ZT2, ZT3, ZT4, ZT5, ZT6, C, D,\
        H, L, L6, D1, H1, L1, D2, H2, L2, D3, H3, L3, D4, H4, L4, D5, H5,\
        L5, MP, MV, TST, D0, L0, DH, LH, D7, D8, VA, VLA, V20, Sc,\
        Sic, SV, Z0, KL, KC, AA, LAMD, QD, PD, MM, QQ, Q, A0, A7,\
        A8, M0, CH, CL, V10, VK0, VK, KE, K2K, KV, KLM, KT, KTR, K10, B0P,\
        B0V, B1, B2, B3, B4, B5, B6, B7, B8, B9, C7, C8, S1V, S2V, S3V, \
        DTZ, R0P, R0V, R1, R2, R3, R4, R5, K1, K2, K3, K4, K5, K6, A1, A2,\
        A3, A5, PH, VH, VL, DT, AK, P2, PL, P1, P, VH1, TV, YV, HF, TEE,\
        U0, Y0, T0, QLV, DYV, DYP, FV, FSV, FLV, ET, EV, ES, EK, EL, EC, \
        EO, UIM, J,  flag, T_list, UP_list, UV_list, SP_list, P_list,\
        P1_list, P2_list, PL_list, QA_list, QPV_list, QE_list,\
        QE2_list, Q11, TZ_list, Z1_list, Z2_list, PP1_list, Q11_list,\
        Q12_list, directtoI, FP, FSP, FLP, WP, YP, UP, P3, AP, AV,\
        QL, WV, UV, F0, FS0, W0, DY0, PPH, KA, QPV, QA, QE, QE2,\
        Q1, Q2, Q3, Q4, Q5, ZZ, QL1, QL2, U12, DY1, U1, T, Y1, S,\
        W1, U22, U2, DY2, Y2, W2, TB, SP, A11, X1, X2, X3, X4, X5, X6, I, \
        Z2, Z1, UVQ, Z, EE, FF, GG, Q12, QP, TZ0, VZ, EZL, TR, I1, TVI, \
        N, I2, TI, TTT, PP1, AS, TS, TTI, YYP,\
        TVR, SR, URM, SB, PMAX, PMIN

        print("State I:")
        state_I_paramates()
        while True:
            if YV > SV:
                TVI = TV
                # GOTO I' state
                II_state_2680()
                break
            computation_of_impact_pressure_4040()
            valve_moving_3780()
            back_oil_moving_3830()
            computation_of_flow_4130()
            high_accumulator_3880()
            low_accumulator_3930()
            ET = ET + P * QI * DT
            EV = EV + X2 * FV * DYV
            EK = EK + X2 * FLV * DYV
            EL = EL + ((P1 - PL) * QL1 + (P2 - PL) * QL2 + (P - PL) * QLV) * DT
            TS = TS + DT
            TV = TV + DT
            I1 = I1 + 1
            T_list.append(TTT + TS + TEE)
```

```
        SP_list.append(YP)
        gen_list()
        print("I:   {:<10.5f} {:<10.5f} {:<10.5f} {:<10.5f} "
        "{:<10.5f} {:<10.5f} {:<10.5f}".format(PH / MM, P1
        / MM, PL / MM, UP, Q * YP, UV, Q * YV))
        if TS >= TST:
            break
    TTT = TR + TB + TI + TST
    print("Ti={:<8s} ms".format(format_value(Q * TI)))
    print("------------------------------")# End of one cycle.
def AI_state():
    global NN, NZ, QI, PA, PLA, K, TT, DTV, PII, TAUP, TAUV, MU,\
    EPSP, EPSV, RHO, ZT1, ZT2, ZT3, ZT4, ZT5, ZT6, C, D, H, L,\
    L6, D1, H1, L1, D2, H2, L2, D3, H3, L3, D4, H4, L4, D5, H5, L5,\
    MP, MV, TST, D0, L0, DH, LH, D7, D8, VA, VLA, V20, Sc, Sic,\
    SV, Z0, KL, KC, AA, LAMD, QD, PD, MM, QQ, Q, A0, A7, A8,\
    M0, CH, CL, V10, VK0, VK, KE, K2K, KV, KLM, KT, KTR, K10, \
    B0P, B0V, B1, B2, B3, B4, B5, B6, B7, B8, B9, C7, C8, S1V, S2V,\
    S3V, DTZ, R0P, R0V, R1, R2, R3, R4, R5, K1, K2, K3, K4, K5, K6,\
    A1, A2, A3, A5, PH, VH, VL, DT, AK, P2, PL, P1, P, VH1, TV, YV,\
    HF, TEE, U0, Y0, T0, QLV, DYV, DYP, FV, FSV, FLV, ET, EV, ES,\
    EK, EL, EC, EO, UIM, J, flag, T_list, UP_list, UV_list, SP_list,\
    P_list, P1_list, P2_list, PL_list, QA_list, QPV_list, QE_list,\
    QE2_list, Q11, TZ_list, Z1_list, Z2_list, PP1_list, \
    Q11_list, Q12_list, directtoI, FP, FSP, FLP, WP, YP, UP, P3,\
    AP, AV, QL, WV, UV, F0, FS0, W0, DY0, PPH, KA, QPV, QA,\
    QE, QE2, Q1, Q2, Q3, Q4, Q5, ZZ, QL1, QL2, U12, DY1,\
    U1, T, Y1, S, W1, U22, U2, DY2, Y2, W2, TB, SP, A11, X1, X2, X3,\
    X4, X5, X6, I, Z2, Z1, UVQ, Z, EE, FF, GG, Q12, QP, TZ0, VZ, EZL,\
    TR, I1, TVI, N, I2, TI, TTT, PP1, AS, TS, TTI, YYP, TVR, SR,\
    URM, SB, PMAX, PMIN, TVD, YVD

    print("state A'")
    state_AI_paramates()
    while True:
        computation_of_return_pressure_3980()
        piston_moving_3710()
        valve_moving_3780()
        back_oil_moving_3830()
        computation_of_flow_4130()
        high_accumulator_3880()
        low_accumulator_3930()
        TR = TR + DT
        TV = TV + DT
        I1 = I1 + 1
        T_list.append(TR + TEE)
        SP_list.append(YP)
        gen_list()
        print("A':   {:<10.5f} {:<10.5f} {:<10.5f} {:<10.5f}"
       " {:<10.5f} {:<10.5f} {:<10.5f}"
        .format(PH / MM, P1 / MM, PL /
        MM, UP, Q * YP, UV, Q * YV))
        if YV >= S:
            T = TR
```

```
        W1 = WV
        U1 = UV
        Y1 = YV
        W2 = WP
        U2 = UP
        Y2 = YP
        computation_of_revision_II_4300()
        computation_of_impact_pressure_4040()
        TR = T
        TV = TV - DT
        UV = U1
        YV = S
        UP = U2
        YP = Y2
        TVI = TV
        T_list[-1] = TR + TEE
        SP_list[-1] = YP
        gen_riversion_list()
        print("A':   {:<10.5f} {:<10.5f} {:<10.5f} {:<10.5f}"
      " {:<10.5f} {:<10.5f} {:<10.5f}"
      .format(PH / MM, P1 / MM, PL /
      MM, UP, Q * YP, UV, Q * YV))
        print("Valve Delay (A' State): TvD={:<8s} ms,
        YvD={:<8s} mm".format(
          format_value(Q * (TV - TVD)),
          format_value(Q * (YV - YVD))
        ))
        break

def AII_state():# state A"
  global NN, NZ, QI, PA, PLA, K, TT, DTV, PII, TAUP, TAUV,\
  MU, EPSP, EPSV, RHO, ZT1, ZT2, ZT3, ZT4, ZT5, ZT6, C, D,\
  H, L, L6, D1, H1, L1, D2, H2, L2, D3, H3, L3, D4, H4, L4, D5,\
  H5, L5, MP, MV, TST, D0, L0, DH, LH, D7, D8, VA, VLA,\
  V20, Sc, Sic, SV, Z0, KL, KC, AA, LAMD, QD, PD, MM, QQ, Q, \
  A0, A7, A8, M0, CH, CL, V10, VK0, VK, KE, K2K, KV, KLM, KT,\
  KTR, K10, B0P, B0V, B1, B2, B3, B4, B5, B6, B7, B8, B9, C7, C8,\
  S1V, S2V, S3V, DTZ, R0P, R0V, R1, R2, R3, R4, R5, K1, K2, K3,\
  K4, K5, K6, A1, A2, A3, A5, PH, VH, VL, DT, AK, P2, PL, P1, P,\
  VH1, TV, YV, HF, TEE, U0, Y0, T0, QLV, DYV, DYP, FV, FSV,\
  FLV, ET, EV, ES, EK, EL, EC, EO, UIM, J, flag, T_list, UP_list,\
  UV_list, SP_list, P_list, P1_list, P2_list, PL_list, QA_list,\
  QPV_list, QE_list, QE2_list, Q11, TZ_list, Z1_list, Z2_list, \
  PP1_list, Q11_list, Q12_list, directtoI, FP, FSP, FLP, WP, YP,\
   UP, P3, AP, AV, QL, WV, UV, F0, FS0, W0, DY0, PPH, KA, QPV,\
   QA, QE, QE2, Q1, Q2, Q3, Q4, Q5, ZZ, QL1, QL2, U12, DY1, U1,\
  T, Y1, S, W1, U22, U2, DY2, Y2, W2, TB, SP, A11, X1, X2,\
  X3, X4, X5, X6, I, Z2, Z1, UVQ, Z, EE, FF, GG, Q12, QP,\
  TZ0, VZ, EZL, TR, I1, TVI, N, I2, TI, TTT, PP1, AS, TS, TTI, YYP,\
  TVR, SR, URM, SB, PMAX, PMIN, TVD, YVD

  print('A"')
  state_AII_paramates()
  while True:
    computation_of_impact_pressure_4040()
```

```
        valve_moving_3780()
        back_oil_moving_3830()
        computation_of_flow_4130()
        high_accumulator_3880()
        low_accumulator_3930()
        computation_of_energy_loss_4170()
        TR = TR + DT
        TV = TV + DT
        I1 = I1 + 1
        T_list.append(TR + TEE)
        SP_list.append(YP)
        gen_list()
        print('A":   {:<10.5f} {:<10.5f} {:<10.5f} {:<10.5f}'
        '{:<10.5f} {:<10.5f} {:<10.5f}'
        .format(PH / MM, P1 / MM, PL /
        MM, UP, Q * YP, UV, Q * TV))
        if YV >= S2V:
          T = TV
          W1 = WV
          U1 = UV
          Y1 = YV
          computation_of_revision_I_4240()
          computation_of_return_pressure_3980()
          if PP1 < 0:
            P1 = P1 + AA * RHO * abs(UIM)
          TV = T
          TR = TR - DT
          UV = U1
          YV = S
          T_list[-1] = TR + TEE
          SP_list[-1] = YP
          gen_riversion_list()
          print('A":   {:<10.5f} {:<10.5f} {:<10.5f} {:<10.5f}'
          '{:<10.5f} {:<10.5f}{:<10.5f}'
          .format(PH / MM,P1 / MM, PL
          / MM, UP,Q * YP, UV, Q * TV))
          print("Valve Delay (A' State): TvD={:<8s}ms,
          YvD={:<8s}mm".format(format_value(Q * (TV - TVD)),
          format_value(Q * (YV - YVD))))
          print("------------------------------")
          break

  def format_value(value):
    if abs(value) >= 1e3 or 0 < abs(value) < 1e-3:
      return "{:8.3e}".format(value)
    else:
      return "{:8.3f}".format(value)

  def output_function(f):

    PT = 0
    P_list_end = P_list[-(I1+1):-1]
    T_list_end = T_list[-(I1+1):-1]

    for i in range(1, I1):
```

```
        PT = PT + P_list_end[i]* (T_list_end[i] - T_list_end[i - 1])
    PAVE = PT /(T_list_end[-1]-T_list_end[0])
    #Calculate the average pressure.
    DEV = MM*((CH/PMIN)**AK - (CH/PMAX)**AK)
    w = 10000
    print("Ei:{:<8s} J  f:{:<8s} Hz"
    .format(format_value(EI), format_value(F)))
    print("Eta:{:<8s}  EtaM:{:<8s}  EtaV:{:<8s}  EtaP:{:<8s}"
    .format(format_value(ETA), format_value(ETAM),
    format_value(ETAV), format_value(ETAP)))
    print("Pmax:{:<8s} MPa  Pmin:{:<8s} MPa  Pave:{:<8s} MPa"
    .format(format_value(PMAX / MM), format_value(PMIN / MM),
    format_value(PAVE / MM)))
    print("Urm:{:<8s} m/s  Uim:{:<8s} m/s  DEVh:{:<8s}"
    .format(format_value(URM),
    format_value(UIM), format_value(DEV)))
    print("Tr:{:<8s} ms  Tb:{:<8s} ms  Ti:{:<8s} ms"
    .format(format_value(TR * Q), format_value(TB * Q),
    format_value(TI * Q)))
    print("Tvr:{:<8s} ms  Tvi:{:<8s} ms".
    format(format_value(TVR * Q), format_value(TVI * Q)))
    print("Sp:{:<8s} mm  Sr:{:<8s} mm  Sb:{:<8s} mm"
    .format(format_value(SP * Q), format_value(SR * Q),
    format_value(SB * Q)))
    print("Sc:{:<8s} mm  Sic:{:<8s} mm"
    .format(format_value(Sc * Q), format_value(Sic * Q)))
    print("Qi:{:<8s} L/min  Pa:{:<8s} MPa  PLa:{:<8s} MPa  K:{:<6s}"
    .format(format_value(QI * QQ), format_value(PA / MM),
    format_value(PLA / MM), format_value(K)))
    print("Mp:{:<8s} kg  Mv:{:<8s} kg  Va:{:<8s} ml  Vla:{:<8s} ml"
    .format(format_value(MP), format_value(MV),
    format_value(MM * VA), format_value(MM * VLA)))
    print("A1:{:<8s} cm^2  A2:{:<8s} cm^2 "
    "A3:{:<8s} cm^2  A5:{:<8s} cm^2"
    .format(format_value(A1 * w), format_value(A2 * w),
    format_value(A3 * w), format_value(A5 * w)))
    print("TauP:{:<8s}  TauV:{:<8s}  EpsP:{:<8s}"
    "EspV:{:<8s}  MU:{:<8s}".format(format_value(TAUP),
    format_value(TAUV), format_value(EPSP),
    format_value(EPSV), format_value(MU)))
    print("Zeta1:{:<8s} Zeta2:{:<8s} Zeta3:{:<8s}"
    "Zeta4:{:<8s} Zeta5:{:<8s} "
    "Zeta6:{:<8s}".format(format_value(ZT1), format_value(ZT2),
    format_value(ZT3), format_value(ZT4),
    format_value(ZT5), format_value(ZT6)))
    print("Et:{:<8s}  Ev:{:<8s}  Es:{:<8s}  Ek:{:<8s}".format(
        format_value(ET), format_value(EV),
         format_value(ES), format_value(EK)))
    print("Eo:{:<8s}  El:{:<8s}  Ec:{:<8s}".format(
        format_value(EO), format_value(EL), format_value(EC)))
    return

if __name__ == '__main__':

    (NN, NZ, QI, PA, PLA, K, TT, DTV, PII, TAUP, TAUV, MU,
```

```
EPSP, EPSV, RHO, ZT1, ZT2, ZT3, ZT4, ZT5, ZT6, C, D, H,
L, L6, D1, H1, L1, D2, H2, L2, D3, H3, L3, D4, H4, L4, D5,
 H5, L5, MP, MV, TST, D0, L0, DH, LH, D7, D8, VA, VLA,
V20, Sc, Sic, SV, Z0, KL, KC, AA, LAMD, QD, PD, MM, QQ,
Q, A0, A7, A8, M0, CH, CL, V10, VK0, VK, KE, K2K, KV,
KLM, KT, KTR, K10, B0P, B0V, B1, B2, B3, B4, B5, B6, B7,
B8, B9, C7, C8, S1V, S2V, S3V, DTZ, R0P, R0V, R1, R2, R3,
R4, R5, K1, K2, K3, K4, K5, K6, A1, A2, A3, A5, PH, VH, VL,
DT, AK, P2, PL, P1, P, VH1, TV, YV, HF, TEE, U0, Y0, T0, QLV,
DYV, DYP, FV, FSV, FLV, ET, EV, ES, EK, EL, EC, EO, UIM, J,
flag, T_list, UP_list, UV_list, SP_list, P_list, P1_list,
P2_list, PL_list, QA_list, QPV_list, QE_list, QE2_list, Q11,
TZ_list, Z1_list, Z2_list, PP1_list, Q11_list, Q12_list,
 directtoI, FP, FSP, FLP, WP, YP, UP, P3, AP, AV, QL, WV,
UV, F0, FS0, W0, DY0, PPH, KA, QPV, QA, QE, QE2, Q1,
Q2, Q3, Q4, Q5, ZZ, QL1, QL2, U12, DY1, U1, T, Y1, S, W1, U22, U2,
DY2, Y2, W2, TB, SP, A11, X1, X2, X3, X4, X5, X6, I, Z2, Z1,
UVQ, Z, EE, FF, GG, Q12, QP, TZ0, VZ, EZL, TR, I1, TVI, N, I2, TI,
TTT, PP1, AS, TS, TTI, YYP, TVR, SR, URM, SB, PMAX, PMIN,
TVD, YVD) = initail()

UP = K * UIM
YP = 0
TR = 0
I1 = 0

P1_list_stable2 = []
P_list_stable2 = []
QE_list_stable2 = []
QPV_list_stable2 = []
SP_list_stable2 = []
UP_list_stable2 = []
QT_list = []
QT_list_stable2=[]
PL_list_stable2=[]
QE2_list_stable2 = []

while True:
  A_state()
  B_state()
  C_state()
  D_state()
  E_state()
  F_state()
  G_state()
  if directtoI:
    I_state()
  else:
    HI_state()#go to H'
    II_state()#go to I'

  VH2 = VH
  J = J + 1
  print("N#", J, "  Vh1", format_value(MM * VH1), "  Vh2",
  format_value(MM * VH2))
```

```
F = 1 / TTT
EI = 0.5 * MP * UIM ** 2
print("F:{:<8s} Hz  Ei:{:<8s} J  Pmax:{:<8s} Mpa "
"Pmin:{:<8s} Mpa  "
    "Sp:{:<8s} mm".format(
  format_value(F),
  format_value(EI),
  format_value(PMAX / MM),
  format_value(PMIN / MM),
  format_value(SP * Q)
))
ETA = 0.5 * MP * (UIM ** 2) / ET
ETAM = 1 - (ES + EK + EO) / ET
ETAV = (A1 - A2) * SP / QI / TTT
ETAP = 1 - EC / ET
print("Number******", J)
print("Eta:{:<8s}  EtaM:{:<8s}  EtaV:{:<8s}  EtaP:{:<8s}"
.format(format_value(ETA), format_value(ETAM),
format_value(ETAV), format_value(ETAP)))
print("ETA3: {:<8s}".format(format_value(ETAM * ETAV
* ETAP)))

print("Et:{:<8s}  Ev:{:<8s}  Es:{:<8s} Mpa  Ek:{:<8s}"
.format(
  format_value(ET), format_value(EV), format_value(ES),
  format_value(EK)))
print("Eo:{:<8s}  El:{:<8s}  Ec:{:<8s} Mpa".format(
  format_value(EO), format_value(EL), format_value(EC)))
DE = ET - EV - ES - EK - EO - EL - EC - ((1 - K) ** 2) * EI
print("DE: {:<8s}".format(format_value(DE)))

print("Tvr:{:<8s} ms  Tvi:{:<8s} ms".format(
  format_value(TVR * Q), format_value(TVI * Q)))
if abs(VH2 - VH1) <= DTV:# Meet the simulation
#termination condition.
  print("I1=", I1)
  output_function(F) # Output the simulation results.
  draw_picture() # Output the simulation curves.
  break

else:
  VH1 = VH2
  ET = 0
  EV = 0
  ES = 0
  EK = 0
  EO = 0
  EL = 0
  EC = 0
  if YV >= SV: #The valve spool stops moving.
    TVI = TV # Record the stroke reversal time.
    print("A")
    UP = K * UIM
    YP = 0
    TR = 0
```

```
            I1 = 0
            TEE += TTT
            continue # Return to state A.
        else:
            TR = 0
            YP = 0
            YVD = YV
            TVD = TV
            print("TVD = ", TV)
            I1 = 0
            TEE += TTT
            if YV >= S2V:#The oil circuit has been switched.
                UP = K * UIM
            else:
                AII_state()#The oil circuit has not been switched;
                        # proceed to state A".
            AI_state()# Enter stae A'
```

4. The Code of Paramaters.py

```
import numpy as np
import pandas as pd

def read_excel_parameters(excel_file):
    """
    Use pandas to read the Excel file and load the parameter
names and values into a dictionary.
    """
   # Use the read_excel function from pandas to read the Excel file.
    df = pd.read_excel(excel_file, header=None)

    parameters = {}

    for index, row in df.iterrows():
        parameter_name = row[0]
        parameter_value = row[1]
        parameter_value = float(parameter_value)
        parameters[parameter_name] = parameter_value

    return parameters

def parameters70(parameters_dict):
    QI = parameters_dict['QI']
    PA = parameters_dict['PA']
    PLA = parameters_dict['PLA']
    K = parameters_dict['K']
    TT = parameters_dict['TT']
    DTV = parameters_dict['DTV']
    return QI, PA, PLA, K, TT, DTV

def parameters90(parameters_dict):
    PII = 3.1416
    TAUP = parameters_dict['TAUP']
    TAUV = parameters_dict['TAUV']
    MU = parameters_dict['MU']
```

```
    EPSP = parameters_dict['EPSP']
    EPSV = parameters_dict['EPSV']
    RHO = parameters_dict['RHO']
    ZT1 = parameters_dict['ZT1']
    ZT2 = parameters_dict['ZT2']
    ZT3 = parameters_dict['ZT3']
    ZT4 = parameters_dict['ZT4']
    ZT5 = parameters_dict['ZT5']
    ZT6 = parameters_dict['ZT6']
    C = parameters_dict['C']
    return PII, TAUP, TAUV, MU, EPSP, EPSV, RHO,\
    ZT1, ZT2, ZT3, ZT4, ZT5, ZT6, C

def parameters110(parameters_dict):
    D = parameters_dict['D']
    H = parameters_dict['H']
    L = parameters_dict['L']
    L6 = parameters_dict['L6']
    D1 = parameters_dict['D1']
    H1 = parameters_dict['H1']
    L1 = parameters_dict['L1']
    D2 = parameters_dict['D2']
    H2 = parameters_dict['H2']
    L2 = parameters_dict['L2']
    return D, H, L, L6, D1, H1, L1, D2, H2, L2

def parameters130(parameters_dict):
    D3 = parameters_dict['D3']
    H3 = parameters_dict['H3']
    L3 = parameters_dict['L3']
    D4 = parameters_dict['D4']
    H4 = parameters_dict['H4']
    L4 = parameters_dict['L4']
    D5 = parameters_dict['D5']
    H5 = parameters_dict['H5']
    L5 = parameters_dict['L5']
    return D3, H3, L3, D4, H4, L4, D5, H5, L5

def parameters150(parameters_dict):
    MP = parameters_dict['MP']
    MV = parameters_dict['MV']
    TST = parameters_dict['TST']
    return MP, MV, TST

def parameters170(parameters_dict):
    D0 = parameters_dict['D0']
    L0 = parameters_dict['L0']
    DH = parameters_dict['DH']
    LH = parameters_dict['LH']
    D7 = parameters_dict['D7']
    D8 = parameters_dict['D8']
    return D0, L0, DH, LH, D7, D8

def parameters190(parameters_dict):
    VA = parameters_dict['VA']
```

```
    VLA = parameters_dict['VLA']
    V20 = parameters_dict['V20']
    return VA, VLA, V20

def parameters210(parameters_dict):
    Sc = parameters_dict['Sc']
    Sic = parameters_dict['Sic']
    SV = parameters_dict['SV']
    Z0 = parameters_dict['Z0']
    return Sc, Sic, SV, Z0

def parameters230(parameters_dict):
    KL = parameters_dict['KL']
    KC = parameters_dict['KC']
    AA = parameters_dict['AA']
    return KL, KC, AA

def parameters250(parameters_dict):
    LAMD = parameters_dict['LAMD']
    QD = parameters_dict['QD']
    PD = parameters_dict['PD']
    return LAMD, QD, PD

def parameters270():
    MM = 1000000
    QQ = 60000
    Q = 1000
    return MM, QQ, Q

def coefficient280(QI, PA, PLA, VA, VLA, MM, QQ):
    QI = QI / QQ
    PA = MM * PA
    PLA = MM * PLA
    VA = VA / MM
    VLA = VLA / MM
    return QI, PA, PLA, VA, VLA

def coefficient290(DTV, V20, MM):
    DTV = DTV / MM
    V20 = V20 / MM
    return DTV, V20

def coefficient300(PII, D0, D7, D8):
    A0 = (PII * D0 ** 2) / 4
    A7 = (PII * D7 ** 2) / 4
    A8 = (PII * D8 ** 2) / 4
    return A0, A7, A8

def coefficient310(RHO, A0, L0, PA, VA, PLA, VLA):
    M0 = RHO * A0 * L0
    CH = PA * VA ** 1.4
    CL = PLA * VLA ** 1.4
    return M0, CH, CL

def coefficient320(PII, DH, LH, V20):
    V10 = PII * DH ** 2 * LH / 4
```

```
    VK0 = V10 + V20
    VK = VK0
    return V10, VK0, VK

def coefficient330(VK0, KC, KL, V20, V10):
    KE = VK0 * KC * KL / (V10 * KL + VK0 * KC)
    K2K = V20 * KC / (V10 * KC + V10 * KL + V20 * KC)
    return KE, K2K

def coefficient340(VK, KE):
    KV = VK / KE
    return KV

def coefficient350(LAMD, RHO, L0, D0, QD, QQ, MM, PD):
    KLM = LAMD * RHO * L0 / D0 / 2
    KT = 1.0 * QD / QQ / ((MM * PD) ** 0.5)
    return KLM, KT

def coefficient360(A0, KT, KLM):
    KTR = A0 * A0 / KT / KT
    K10 = A0 * (KLM + KTR)
    return KTR, K10

def coefficient370(PII, MU, EPSP, EPSV):
    B0P = PII * MU / ((1 - EPSP ** 2) ** 0.5)
    B0V = PII * MU / ((1 - EPSV ** 2) ** 0.5)
    return B0P, B0V

def coefficient380_420(B0P, D, L, H, L6, D1, L1, H1, D2, L2, H2, TAUP,
B0V, D3, L3, H3, D4, L4, H4, D5, L5, H5, TAUV, PII):
    B1 = B0P * (D * L / H + D * L6 / H + D1 * L1 / H1 + D2 * L2 / H2)
    B2 = PII * (D2 * H2 - D * H) / 2
    B3 = PII * (D * H - D1 * H1) / 2
    B4 = TAUP * (D * L + D2 * L2)
    B5 = TAUP * (D * L6 + D1 * L1)
    B6 = B0V * (D3 * L3 / H3 + 2 * D4 * L4 / H4 + D5 * L5 / H5)
    B7 = PII * D3 * H3 / 2
    B8 = TAUV * (D3 * L3 + 2 * D4 * L4)
    B9 = TAUV * D5 * L5
    return B1, B2, B3, B4, B5, B6, B7, B8, B9

def coefficient430(C, PII, D4, RHO):
    C7 = C * PII * D4 * (2 / RHO) ** 0.5
    C8 = 1 / C7 ** 2
    return C7, C8

def coefficient440(SV, Z0, NZ):
    S1V = (SV - Z0) / 2
    S2V = S1V + Z0
    S3V = SV / 2
    DTZ = Z0 / NZ
    return S1V, S2V, S3V, DTZ

def coefficient450_480(PII, EPSP, MU, EPSV, D, H, L6, D1, H1, L1, L,
D2, H2, L2, D3, H3, L3, D4, H4, L4, D5, H5, L5):
    R0P = PII * (1 + 1.5 * EPSP ** 2) / 12 / MU
```

```
    R0V = PII * (1 + 1.5 * EPSV ** 2)
    R1 = R0P * (D * H ** 3 / L6 + D1 * H1 ** 3 / L1)
    R2 = PII * (D1 * H1 - D * H) / 2
    R3 = R0P * (D * H ** 3 / L + D2 * H2 ** 3 / L2)
    R4 = PII * (D * H - D2 * H2) / 2
    R5 = R0V * (D3 * H3 ** 3 / L3 + 2 * D4 * H4 ** 3 / L4 + D5 *
    H5 ** 3 / L5)
    return R0P, R0V, R1, R2, R3, R4, R5

def coefficient490_500(ZT1, ZT2, ZT3, ZT4, ZT5, ZT6, RHO, A7, A8):
    K1 = ZT1 * RHO / A8 ** 2 / 2
    K2 = ZT2 * RHO / A8 ** 2 / 2
    K3 = ZT3 * RHO / A8 ** 2 / 2
    K4 = ZT4 * RHO / A8 ** 2 / 2
    K5 = ZT5 * RHO / A7 ** 2 / 2
    K6 = ZT6 * RHO / A8 ** 2 / 2
    return K1, K2, K3, K4, K5, K6

def coefficient510_520(PII, D, H, D1, H1, D2, H2, D3, H3, D5, H5):
    A1 = PII * ((D - 2 * H) ** 2 - (D1 - 2 * H1) ** 2) / 4
    A2 = PII * ((D - 2 * H) ** 2 - (D2 - 2 * H2) ** 2) / 4
    A3 = PII * (D3 - 2 * H3) ** 2 / 4
    A5 = PII * (D5 - 2 * H5) ** 2 / 4
    return A1, A2, A3, A5

def coefficient530_540(PA,VA,VLA,TT,PLA):
    PH = PA
    VH = VA
    VL = VLA
    DT = TT
    AK = 1 / 1.4
    P2 = PH
    PL = PLA
    P1 = PL
    P = PH
    VH1 = VA
    return PH, VH, VL, DT, AK, P2, PL, P1, P, VH1

def initialize():
    TV = 0
    YV = 0
    HF = 0
    TEE = 0
    U0 = 0
    Y0 = 0
    T0 = 0
    QLV = 0
    DYV = 0
    DYP = 0
    FV = 0
    FSV = 0
    FLV = 0
    ET = 0
    EV = 0
    ES = 0
```

```
EK = 0
EL = 0
EC = 0
EO = 0
UIM = 0
J = 0
flag = 0
Q11 = 0
TZ_list = []
Z1_list = []
Z2_list = []
PP1_list = []
Q11_list = []
Q12_list = []

T_list = []
UP_list = []
UV_list = []
SP_list = []
P_list = []
P1_list = []
P2_list = []
PL_list = []
QA_list = []
QPV_list = []
QE_list = []
QE2_list = []
return TV, YV, HF, TEE, U0, Y0, T0, QLV, DYV, DYP, FV, FSV, \
    FLV, ET, EV, ES, EK, EL, EC, EO, UIM, J, flag,T_list,\
    UP_list, UV_list, SP_list, P_list, P1_list, P2_list, \
    PL_list, QA_list, QPV_list, QE_list, QE2_list, Q11, \
    TZ_list, Z1_list, Z2_list, PP1_list, Q11_list, Q12_list
```

Appendix C
Computer Simulation Program for Hydraulic Impact Mechanisms Without a High-Pressure Accumulator

This program is written in Python. It is primarily designed for computer simulations of multi-state and multi-condition hydraulic impact mechanisms, similar to the YYG250 rear-chamber control type described in the book, but without a high-pressure accumulator.

Similar to Appendix B, this program consists of three parts: the parameter settings table (.xlsx file), parameter assignment (parameters2.py), and the main program. Since there is no high-pressure accumulator, the parameter settings table does not include PA and VA but includes the initial working pressure PS (MPa) as an input. The usage requirements for this program are the same as those described in Appendix B.

1. **Main Program Code**

```
import numpy as np
import paramaters2 as pm
import matplotlib.pyplot as plt

def PlotDemo1(L1, L2):
  global P1_list, QE_list, P_list, QPV_list,PL_list,\
  QE2_list, QT_list,QPV_list_ni
  fig = plt.figure(figsize = (12, 8))
  ax = fig.add_subplot(1, 1, 1)
  color = 'tab:blue'
  ax.plot(L1, L2, color=color)
  if L2 == P1_list:
    ax.set_title("Rear Chamber Pressure
    P1 without High-Accumulator")
    ax.set_xlabel('Time (s)')
    ax.set_ylabel('Rear chamber pressure pressure (MPa)')
    plt.savefig("P1 without High-Accumulator.png")
  elif L2 == QE_list:
    ax.set_title("Compensation Flow Rate Qe")
    ax.set_xlabel('Time (s)')
    ax.set_ylabel('Compensation flow rate Qe (L/min)')
```


Q. He, *Research and Design of Hydraulic Impact Mechanisms*,
https://doi.org/10.1007/978-981-96-4864-1

```
        plt.savefig("Compensation flow rate Qe.png")
    elif L2 == QPV_list_ni:
        ax.set_title("Required Flow Qpv without High-Accumulator")
        ax.set_xlabel('Time (s)')
        ax.set_ylabel('Required Flow Qpv(L/min)')
        plt.savefig("Flow Qpv without high-accumulator.png")
    elif L2 == P_list:
        ax.set_title("Inlet Pressure P without High-Accumulator")
        ax.set_xlabel('Time (s)')
        ax.set_ylabel('Inlet Pressure P (MPa)')
        plt.savefig("pressure P without high-accumulator.png")
    elif L2 == UP_list:
        ax.set_title("Up of Piston without High-Accumulator")
        ax.set_xlabel('Time (s)')
        ax.set_ylabel('Velocity up(m/s)')
        plt.savefig("Velocity Up without high-accumulator.png")
    elif L2 == SP_list:
        ax.set_title("Displacement of Piston Sp"
        "without High-Accumulator")
        ax.set_xlabel('Time (s)')
        ax.set_ylabel('Displacement (mm)')
        plt.savefig("Displacement without high-accumulator.png")
    elif L2 == PL_list:
        ax.set_title("Return Pressure PL")
        ax.set_xlabel('Time (s)')
        ax.set_ylabel('Return pressure PL (MPa)')
        plt.savefig("Return pressure PL.png")
    elif L2 == QE2_list:
        ax.set_title("Compensation Flow Rate Qe2")
        ax.set_xlabel('Time (s)')
        ax.set_ylabel('Compensation flow rate Qe2 (L/min)')
        plt.savefig("Compensation flow rate Qe2.png")
    elif L2 == QT_list:
        ax.set_title("Inlet Flow Rate Qt")
        ax.set_xlabel('Time (s)')
        ax.set_ylabel('Inlet flow rate Qt (L/min)')
        plt.savefig("Inlet flow rate Qt.png")
    plt.show()

def PlotDemo_SP_UP_together():
    figure, ax1 = plt.subplots(figsize=(12, 8))
    ax1.set_title("Velocity and Displacement of
    Piston without High-Accumulator")
    color = 'tab:red'
    ax1.set_xlabel('Time (s)')
    ax1.set_ylabel('Velocity (m/s)', color=color)
    ax1.plot(T_list, UP_list, color=color)
    ax1.tick_params(axis='y', labelcolor=color)

    ax2 = ax1.twinx()
    color = 'tab:blue'
    ax2.set_ylabel('Displacement Sp(mm)', color=color)
    ax2.plot(T_list, SP_list, color=color)
    ax2.tick_params(axis='y', labelcolor=color)
    figure.tight_layout()
```

```
    plt.savefig("Velocity and displacement of piston"
     "without high-accumulator.png")
    plt.show()

def PlotDemo_Stable(L1, L2):
   # Display and save the curve plot after the simulation stabilizes
    global P1_list_stable2, P_list_stable2, QE_list_stable2,\
     QPV_list_stable2, SP_list_stable2, UP_list_stable2,\
     QE2_list_stable2,PL_list_stable2,\
      QT_list_stable2
    fig = plt.figure(figsize=(12, 8))
    ax = fig.add_subplot(1, 1, 1)
    color = 'tab:blue'
    ax.plot(L1, L2, color=color)
    if L2 == P1_list_stable2:
        ax.set_title("Rear Chamber Pressure of Piston P1"
        "without high-accumulate")
        ax.set_xlabel('Time (s)')
        ax.set_ylabel('Rear chamber pressure P1 (MPa)')
        plt.savefig("Stabilized P1 without"
         "high-accumulate.png")
    elif L2 == QE_list_stable2:
        ax.set_title("Compensation Flow Rate Qe"
         "without high-accumulate")
        ax.set_xlabel('Time (s)')
        ax.set_ylabel('Compensation flow rate Qe (L/min)')
        plt.savefig("Stabilized Qe without high-accumulate.png")
    elif L2 == QPV_list_stable2:
        ax.set_title("Flow Rate Qpv without high-accumulate")
        ax.set_xlabel('Time (s)')
        ax.set_ylabel('Flow rate required Qpv (L/min)')
        plt.savefig("Stabilized Qpv without high-accumulate.png")
    elif L2 == P_list_stable2:
        ax.set_title("Inlet Pressure P without high-accumulate")
        ax.set_xlabel('Time (s)')
        ax.set_ylabel('Inlet Pressure P (MPa)')
        plt.savefig("Stabilized P without high-accumulate.png")
    elif L2 == UP_list_stable2:
       ax.set_title("Velocity of Piston Up without high-accumulate")
        ax.set_xlabel('Time (s)')
        ax.set_ylabel('Velocity (m/s)')
        plt.savefig("Stabilized Up without high-accumulate.png")
    elif L2 == SP_list_stable2:
        ax.set_title("Displacement of Piston Sp without"
         "high-accumulate")
        ax.set_xlabel('Time (s)')
        ax.set_ylabel('Displacement (mm)')
        plt.savefig("Stabilized Sp without high-accumulate.png")
    elif L2 == QE2_list_stable2:
        ax.set_title("Compensation Flow Rate QE2")
        ax.set_xlabel('Time (s)')
        ax.set_ylabel('QE2 Flow (L/min)')
        plt.savefig("Stabilized QE2 without high-accumulate.png")
    elif L2==QT_list_stable2:
```

```
        ax.set_title("Inlet Flow Rate Qt without high-accumulate")
        ax.set_xlabel('Time (s)')
        ax.set_ylabel('inlet flow rate Qt (L/min)')
        plt.savefig("Stabilized Inlet Flow Rate Qt.png")
    elif L2==PL_list_stable2:
        ax.set_title("return pressure PL without high-accumulate")
        ax.set_xlabel('Time (s)')
        ax.set_ylabel('return pressure PL (PMa)')
        plt.savefig("Stabilized PL without high-accumulate.png")
    plt.show()

def draw_picture():
    global P1_list_stable2, P_list_stable2, QE_list_stable2,\
     QPV_list_stable2, SP_list_stable2, UP_list_stable2, QT_list,\
     QE2_list_stable2,QT_list_stable2,\
     PL_list_stable2,QPV_list_ni
    """Calculate QT_list"""
    QT_list = [QI*QQ + a - b for a, b in zip(QE_list, QE2_list)]

    QPV_list_ni = [-x for x in QPV_list]

    """Extract the stabilized data, I1"""
    P1_list_stable = P1_list[-I1:]
    P_list_stable = P_list[-I1:]
    QE_list_stable = QE_list[-I1:]
    QPV_list_stable = QPV_list_ni[-I1:]
    SP_list_stable = SP_list[-I1:]
    UP_list_stable = UP_list[-I1:]
    QE2_list_stable = QE2_list[-I1:]

    """Calculate the time values of the curve after stabilization"""
    T_list_stable = T_list[-I1:]
    T_list_stable1 = [a - T_list_stable[0] for a in T_list_stable]
   T_list_stable2 = [a + T_list_stable1[-1] for a in T_list_stable1]
   T_list_stable3 = [a + T_list_stable2[-1] for a in T_list_stable1]
    T_list_stable0 = T_list_stable1 + T_list_stable2\
     + T_list_stable3
    """Extract three cycles of the curve after the simulation stabi-
lizes"""
    P1_list_stable2 = P1_list_stable + P1_list_stable\
     + P1_list_stable
    P_list_stable2 = P_list_stable + P_list_stable\
     + P_list_stable
    QE_list_stable2 = QE_list_stable + QE_list_stable\
    + QE_list_stable
    QPV_list_stable2 = QPV_list_stable + QPV_list_stable\
     + QPV_list_stable
    SP_list_stable2 = SP_list_stable + SP_list_stable \
     + SP_list_stable
    UP_list_stable2 = UP_list_stable + UP_list_stable\
     + UP_list_stable
    QE2_list_stable2 = QE2_list_stable + QE2_list_stable\
     +QE2_list_stable
    PL_list_stable2 = PL_list[-I1:] + PL_list[-I1:]\
     + PL_list[-I1:]
```

```
QT_list_stable2 = QT_list[-I1:] + QT_list[-I1:]\
 + QT_list[-I1:]

print("'
        *******      Welcome to the menu      *******
   **** Save the velocity curve, please enter '1'                  ****
    **** Save the rear chamber pressure curve
    , please enter '2'        ****
    **** Save the inlet pressure curve,
     please enter '3'              ****
    **** Save the compensation flow curve,
     please enter '4'            ****
    **** Save the curve of the flow required,
    please enter '5'           ****
    **** Save all curves from the entire simulation
    process, enter '6'   ****
    **** Save the curves after the simulation stabilizes,
     please enter '7'****
    **** End the main program,
     please enter 'Q' or other key           ****
        "')
while True:
   flag = input("Please select the figure you want: ")
   if flag == 'Q':
      break
   elif flag == '6':
      PlotDemo_SP_UP_together()
      PlotDemo1(T_list, P1_list)
      PlotDemo1(T_list, QE_list)
      PlotDemo1(T_list, P_list)
      PlotDemo1(T_list, QPV_list_ni)
      PlotDemo1(T_list, PL_list)
      PlotDemo1(T_list, QE2_list)
      PlotDemo1(T_list, QT_list)
      PlotDemo1(T_list, P1_list)
      PlotDemo1(T_list, SP_list)
   elif flag == '1':
      PlotDemo1(T_list, UP_list)
   elif flag == '2':
      PlotDemo1(T_list, P1_list)
   elif flag == '3':
      PlotDemo1(T_list, P_list)
   elif flag == '4':
      PlotDemo1(T_list, QE_list)
   elif flag == '5':
      PlotDemo1(T_list, QPV_list_ni)
   elif flag == '7':
      PlotDemo_Stable(T_list_stable0, P1_list_stable2)
      PlotDemo_Stable(T_list_stable0, P_list_stable2)
      PlotDemo_Stable(T_list_stable0, QE_list_stable2)
      PlotDemo_Stable(T_list_stable0, QPV_list_stable2)
      PlotDemo_Stable(T_list_stable0, SP_list_stable2)
      PlotDemo_Stable(T_list_stable0, UP_list_stable2)
      PlotDemo_Stable(T_list_stable0, QE2_list_stable2)
```

```
            PlotDemo_Stable(T_list_stable0, PL_list_stable2)
            PlotDemo_Stable(T_list_stable0, QT_list_stable2)
        else:
            break

def initail():
    excel_file = input("Please enter the name and path of "
                       "the parameter settings file (.xlsx)")
    parameters_dict = pm.read_excel_parameters(excel_file)
    NN = 600
    QI, PS, PLA, K, TT, DTV = pm.parameters60(parameters_dict)
    PII, TAUP, TAUV, MU, EPSP, EPSV, RHO, ZT1, ZT2,\
    ZT3, ZT4, ZT5, ZT6, C = pm.parameters80(parameters_dict)
    D, H, L, L6, D1, H1, L1, D2, H2, L2 =\
    pm.parameters100(parameters_dict)
    D3, H3, L3, D4, H4, L4, D5, H5, L5 =\
    pm.parameters120(parameters_dict)
    MP, MV, TST = pm.parameters140(parameters_dict)
    D0, L0, DH, LH, D7, D8 = pm.parameters160(parameters_dict)
    VLA, V20 = pm.parameters180(parameters_dict)
    Sc, Sic, SV, Z0 = pm.parameters200(parameters_dict)
    KL, KC, AA = pm.parameters220(parameters_dict)
    LAMD, QD, PD = pm.parameters240(parameters_dict)
    MM, QQ, Q = pm.parameters260()

    (QI, PS, PLA, VLA, DTV, V20, A0, A7, A8, M0, CL,
     VK0, V10, KE, K2K, KLM, KT, KTR, K10, B0P,
     B0V, B1, B2, B3, B4, B5, B6, B7, B8, B9, C7, C8, S1V,
     S2V, S3V, DTZ, R0P, R0V, R1, R2, R3, R4, R5, K1, K2,
     K3, K4, K5, K6, A1, A2, A3, A5, DT, VL, VK, P, P2, PL,
     P1, VK1) = pm.coefficient(QI, PS, PLA, TT, DTV, PII,
     TAUP, TAUV, MU, EPSP, EPSV, RHO, ZT1, ZT2, ZT3,
     ZT4, ZT5, ZT6, C, D, H, L, L6, D1, H1, L1, D2, H2, L2, D3,
     H3, L3, D4, H4, L4, D5, H5, L5, D0, L0, DH, LH, D7, D8, VLA,
     V20, SV, Z0, KL, KC, LAMD, QD, PD, MM, QQ)

    (QE, U0, Y0, T0, QLV, DYV, DYP, TEE, FV, FSV, FLV, ET,
     EV, ES, EK, EL, EC, EO, UIM, J, HF, T_list, UP_list,
     UV_list, SP_list, P_list, P1_list, P2_list, PL_list,
     QA_list, QPV_list, QE_list, QE2_list) = pm.initialize()

    print("B1:{:<20.7f}B2:{:<20.12f}B3:{:<20.13f}"
    "B4:{:<20.13f}B5:{:<20.13f}".format(B1, B2, B3, B4, B5))
    print("B6:{:<20.10f}B7:{:<20.10f}B8:{:<20.10f}"
    "B9:{:<20.10f}".format(B6, B7, B8, B9))
    print("R1:{:<20.17f}R2:{:<20.16f}R3:{:<20.17f}"
    "R4:{:<20.16f}R5:{:<20.16f}".format(R1, R2, R3, R4, R5))
    print("K1:{:<20.7f}K2:{:<20.7f}K3:{:<20.7f}K4:{:<20.7f}"
    "K5:{:<20.7f}K6:{:<20.7f}".format(K1, K2, K3, K4, K5, K6))
    print("A1:{:<20.11f}A2:{:<20.11f}A3:{:<20.11f}"
    "A5{:<20.11f}".format(A1, A2, A3, A5))

    FP, FSP, FLP, WP, YP, UP = 0, 0, 0, 0, 0, 0
    WV, YV, UV, P3 = 0, 0, 0, 0
    F0, FS0, W0, DY0 = 0, 0, 0, 0
```

```
    QPV, DVK, AP, AV, QL, QE2 = 0, 0, 0, 0, 0, 0
    A11 = 0
    Q1, Q2, Q3, Q5, QT, ZZ, X1, X2 = 0, 0, 0, 0, 0, 0, 0, 0
    X3, X4, Q11 = 0, 0, 0
    X5 = 0
    X6 = 0
    U12, DY1, U1, T, Y1, S, W1 = 0, 0, 0, 0, 0, 0, 0
    U22, U2, DY2, Y2, W2 = 0, 0, 0, 0, 0
    TB, SP = 0, 0
    I, Z1, Z2, UVQ, EE, FF, GG, Q12 = 0, 0, 0, 0, 0, 0, 0, 0
    TVI, TV, SVV = 0, 0, 0
    N = 0
    TI, TTT, TR = 0, 0, 0
    KK, PP1, AS = 0, 0, "G"
    TTI, YYP = 0, 0
    TS = 0
    I1 = 0
    SR, URM = 0, 0
    ELZ = 0
    SB = 0
    PMAX, VKMAX, directtoI, PMIN = 0, 0, False, 0
    TVD, YVD = 0, 0
    TVR = 0
    return (NN, QI, PS, PLA, K, TT, DTV, PII, TAUP, TAUV, MU, EPSP,
     EPSV, RHO, ZT1, ZT2, ZT3, ZT4, ZT5, ZT6, C, D, H, L, L6, D1,
     H1, L1, D2, H2, L2, D3, H3, L3, D4, H4, L4, D5, H5, L5, MP,
     MV, TST, D0, L0, DH, LH, D7, D8, VLA, V20, Sc, Sic, SV, Z0,
     KL, KC, AA, LAMD, QD, PD, MM, QQ, Q, A0, A7, A8, M0, CL, VK0,
     V10, KE, K2K, KLM, KT, KTR, K10, B0P, B0V, B1, B2, B3, B4, B5,
     B6, B7, B8, B9, C7, C8, S1V, S2V, S3V, DTZ, R0P, R0V, R1, R2,
     R3, R4, R5, K1, K2, K3, K4, K5, K6, A1, A2, A3, A5, DT, VL,
     VK, P, P2, PL, P1, VK1,QE, U0, Y0, T0, QLV, DYV, DYP, TEE, FV,
     FSV, FLV, ET, EV, ES, EK, EL, EC, EO, UIM, J, HF, T_list,
     UP_list, UV_list, SP_list, P_list, P1_list, P2_list, PL_list,
     QA_list, QPV_list, QE_list, QE2_list, FP, FSP, FLP, WP, YP,
     UP, WV, YV, UV, P3, F0, FS0, W0, DY0, QPV, DVK, AP, AV, QL,
     QE2, A11, Q1, Q2, Q3, Q5, QT, ZZ, X1, X2, X3, X4, Q11, X5,
     X6, U12, DY1, U1, T, Y1, S, W1, U22, U2, DY2, Y2, W2, TB,
     SP, I, Z1, Z2, UVQ, EE, FF, GG, Q12, TVI, TV, SVV, N, TI,
     TTT, TR, KK, PP1, AS, TTI, YYP, TS, I1, SR, URM, ELZ, SB,
     PMAX, VKMAX, directtoI, PMIN, TVD, YVD, TVR)

def piston_moving_4020():
    global FP, FSP, FLP, WP, DYP, YP, UP
    FP = A2 * P2 - A1 * P1
    FSP = B1 * UP - np.sign(UP) * (B2 * (P2 - PL) + B3 * (P1 - PL))
    FLP = np.sign(UP) * (B4 * (P2 - PL) + B5 * (P1 - PL))
    WP = (FP - FSP - FLP) / MP
    DYP = UP * DT + WP * DT ** 2 / 2
    YP = YP + DYP
    UP = UP + WP * DT

def valve_moving_4090():
    global FV, FSV, FLV, WV, DYV, YV, UV
```

```
    FV = abs(A3 * (P - PL) - A5 * (P - P3))
    FSV = B6 * UV - B7 * (P - PL)
    FLV = B8 * (P - PL) + B9 * (P - P3)
    WV = (FV - FSV - FLV) / MV
    DYV = UV * DT + (WV * DT ** 2) / 2
    YV = YV + DYV
    UV = UV + WV * DT

def backoil_moving_4140():
    global F0, FS0, W0, DY0, Y0, U0, T0
    F0 = A0 * PL
    FS0 = K10 * U0 ** 2
    W0 = (F0 - FS0) / M0
    DY0 = U0 * DT + W0 * DT ** 2 / 2
    Y0 = Y0 + DY0
    U0 = U0 + W0 * DT
    T0 = T0 + DT

def high_accumulator_4190():
    global QPV, DVK, VK, P, QE, QE2
    QPV = (AP * DYP + AV * DYV) / DT
    DVK = (QPV + QL - QI) * DT
    VK = VK - DVK
    if VK < VK0:
        VK = VK0
    P = (VK / VK0 - 1) * KE
    QE = DVK / DT
    QE2 = K2K * QE

def low_accumulator_4240():
    global VL, PL
    VL = VL + A0 * DY0 - A11 * DYP - QL * DT
    if VL > VLA:
        VL = VLA
    PL = CL / VL ** 1.4

def DPr_4290():
    global Q1, Q2, Q3, Q5, QT, P2, ZZ, P1, P3

    Q1 = A1 * UP
    Q2 = A2 * UP
    Q3 = QI + QE
    Q5 = A5 * UV
    QT = QI + QE
    P2 = P - (K2 + K3) * Q2 ** 2 - K4 * QT ** 2
    ZZ = abs(S2V - YV)
   # if ZZ < Z0:
   #     ZZ = Z0
    P1 = PL + K6 * Q1 ** 2 + X2 * C8 * (Q1 / ZZ) ** 2
    P3 = X1 * PL + X2 * (P2 - K5 * Q5 ** 2)

def DPi_4350():
    global Q1, Q2, Q3, QT, P2, ZZ, P1, P3, X3, X4, Q11

    Q1 = A1 * UP
    Q2 = A2 * UP
```

```
    Q3 = QI + Q1 + QE - X3 * Q11
    QT = QI + QE
    P2 = P - np.sign(Q2) * (K2 + K3) * Q2 ** 2
    ZZ = abs(S1V - YV)
    if ZZ < Z0:
        ZZ = 3 * Z0
    P1 = X4 * (P + np.sign(Q1) * K1 * Q1 ** 2 + X2 * np.sign(UP)
            * C8 * (Q1 / ZZ) ** 2 - K4 * QT ** 2)
    P3 = X1 * PL + X2 * (P2 - K5 * Q5 ** 2)
    if P1 < 0:
        P1 = 0

def computation_flow_4430():
    global QL1, QL2, QLV, QL
    QL1 = X5 * (R1 * (P1 - PL) + np.sign(UP) * R2 * UP)
    QL2 = R3 * (P2 - PL) + np.sign(UP) * R4 * UP
    QLV = R5 * (P - PL)
    QL = QL1 + QL2 + QLV

def computation_EnergyLoss_4470():
    global ET, EV, ES, EK, EO, EL, ZZ, EC
    ET = ET + P * QI * DT
    EV = EV + X2 * FV * DYV
    ES = ES + FSP * DYP + X2 * FSV * DYV
    EK = EK + abs(FLP * DYP) + X2 * FLV * DYV
    EO = EO + X6 * A1 * PL * abs(DYP)
    EL = EL + ((P1 - PL) * QL1 + (P2 - PL) * QL2 + (P - PL) * QLV) * DT
    ZZ = abs(S3V - YV)
    if ZZ < Z0:
        ZZ = Z0
    EC = (EC + (X5 * K1 + C8 / ZZ ** 2) * abs(Q1 ** 3 * DT) +
        (K2 * abs(Q2 ** 3) + K3 * abs(Q3 ** 3) + K4 * abs(QT ** 3) +
         X2 * K5 * abs(Q5 ** 3)) * DT)

def computation_revisionI_4540():
    global U12, DY1, U1, DT, T, Y1, DYP, DYV, QPV, DVK, VK, P, QE, QE2
    U12 = U1
    DY1 = Y1 - S
    U1 = np.sign(U1) * (U12 ** 2 - 2 * abs(W1 * DY1)) ** 0.5
    DT = (U12 - U1) / W1
    T = T - DT
    Y1 = S
    if W1 == WP:
        DYP = DY1
    else:
        DYV = DY1
    QPV = (AP * DYP + AV * DYV) / DT
    DVK = (QPV + QL - QI) * DT
    VK = VK + DVK
    P = (VK / VK0 - 1) * KE
    QE = DVK / DT
    QE2 = K2K * QE

def computation_revisionII_4610():
    global U12, DY1, U1, DT, T, Y1, U22, U2, DY2, Y2,\
```

```
    DYP, DYV, QPV, DVK, VK, P, QE, QE2
  U12 = U1
  DY1 = Y1 - S
  U1 = np.sign(U1) * (U12 ** 2 - 2 * abs(W1 * DY1)) ** 0.5
  DT = abs((U12 - U1) / W1)
  T = T - DT
  Y1 = S
  U22 = U2
  U2 = U22 - W2 * DT
  DY2 = (U22 * U22 - U2 * U2) / W2 / 2
  Y2 = Y2 - DY2
  if W1 == WP:
    DYP = DY1
    DYV = DY2
  else:
    DYP = DY2
    DYV = DY1
  QPV = (AP * DYP + AV * DYV) / DT
  DVK = (QPV + QL - QI) * DT
  VK = VK + DVK
  P = (VK / VK0 - 1) * KE
  QE = DVK / DT
  QE2 = K2K * QE

def computation_revisionIII_4690():
  global DT, DYP, TB, YP, UP, SP, QPV, DVK, VK, P, QE, QE2
  DT = UP / WP
  DYP = WP * DT * DT / 2
  TB = TB - DT
  YP = YP - DYP
  UP = 0
  SP = YP
  QPV = (AP * DYP + AV * DYV) / DT
  DVK = (QPV + QL - QI) * DT
  VK = VK + DVK
  P = (VK / VK0 - 1) * KE
  QE = DVK / DT
  QE2 = K2K * QE

def positive_mouth_interval_4750():
  global I, Z1, Z2, UVQ, UV, DT, EE, FF, GG, P1, Q11, Q12
  I = 1
  Z1 = Z1 + DTZ
  Z2 = Z0 - Z1
  UVQ = UV
  UV = (2 * WV * DTZ + UV * UV) ** 0.5
  DT = (UV - UVQ) / WV
  if Z2 == 0:
    Z2 = 0.000000002
  EE = (Z1 * Z1 + I * Z2 * Z2) / Z1 / Z2 / 2
  FF = (C8 * Q1 * Q1 + I * Z2 * Z2 * P) / Z1 / Z2 / 2
  GG = P * P - 4 * I * EE * FF * P + 4 * FF * FF
  while (GG < 0):
    I = -1
    EE = (Z1 * Z1 + I * Z2 * Z2) / Z1 / Z2 / 2
```

```
        FF = (C8 * Q1 * Q1 + I * Z2 * Z2 * P) / Z1 / Z2 / 2
        GG = P * P - 4 * I * EE * FF * P + 4 * FF * FF
    P1 = (2 * EE * FF - I * P - I * GG ** 0.5) / (EE * EE - I) / 2
    Q11 = C7 * Z1 * P ** 0.5
    Q12 = np.sign(P1 - P) * C7 * Z2 * (abs(P - P1)) ** 0.5

def state_A_paramates():
    global AP, AV, A11, X1, X2, X3, X4, X5, X6, \
        TVI, TV, UV, WV, DT, S, SVV, ET, EV, ES, EO, EC, EL
    AP = A2
    AV = 0
    A11 = A1
    X1 = 1
    X2 = 0
    X3 = 0
    X4 = 1
    X5 = 0
    X6 = 1
    TVI = TV
    TV = 0
    UV = 0
    WV = 0
    DT = TT
    S = Sc
    SVV = S2V
    ET = 0
    EV = 0
    ES = 0
    EO = 0
    EC = 0
    EL = 0

def state_B_paramates():
    global AP, AV, A11, X1, X2, X3, X4, X5, X6, \
        DT, TV, UV, YV, TB, N, S
    AP = A2
    AV = A3
    A11 = A1
    X1 = 0
    X2 = 1
    X3 = 0
    X4 = 1
    X5 = 0
    X6 = 1
    DT = TT
    TV = 0
    UV = 0
    YV = 0
    TB = 0
    N = 1
    S = S1V

def state_C_paramates():
    global AP, AV, A11, X1, X2, X3, X4, Z1, Z2, SVV
    AP = A2 - A1
```

```
    AV = A3
    A11 = 0
    X1 = 0
    X2 = 1
    X3 = 1
    X4 = 0
    Z1 = 0
    Z2 = 0
    SVV = S2V

def state_D_paramates():
    global AP, AV, A11, X1, X2, X3, X4, X5, X6, DT, S
    AP = A2 - A1
    AV = A3
    A11 = 0
    X1 = 0
    X2 = 1
    X3 = 0
    X4 = 1
    X5 = 1
    X6 = 0
    DT = TT
    S = SV

def state_E_paramates():
    global AP, AV, A11, X1, X2, X3, X4, X5, X6, DT, TV, UV, WV
    AP = A2 - A1
    AV = 0
    A11 = 0
    X1 = 1
    X2 = 0
    X3 = 0
    X4 = 1
    X5 = 1
    X6 = 0
    DT = TT
    TV = 0
    UV = 0
    WV = 0

def state_F_paramates():
    global AP, AV, A11, X1, X2, X3, X4, X5, X6, TI, YP, DT, S, N, TTT
    AP = A2 - A1
    AV = 0
    A11 = 0
    X1 = 1
    X2 = 0
    X3 = 0
    X4 = 1
    X5 = 1
    X6 = 0
    TI = 0
    YP = 0
    DT = TT
    S = Sic - SP
```

```
    N = 0
    TTT = TR + TB

def state_G_paramates():
    global AP, AV, A11, X1, X2, X3, X4, X5, X6, TV, UV, YV, DT, S,\
        KK, N, PP1, AS
    AP = A2 - A1
    AV = A5 - A3
    A11 = 0
    X1 = 0
    X2 = 1
    X3 = 0
    X4 = 1
    X5 = 1
    X6 = 0
    TV = 0
    UV = 0
    YV = 0
    DT = TT
    S = -SP
    KK = 1
    N = 1
    PP1 = -1
    AS = "G"

def state_H_paramates():
    global T, W1, U1, Y1, S, W2, U2, Y2
    T = TI
    W1 = WV
    U1 = UV
    Y1 = YV
    S = SV
    W2 = WP
    U2 = UP
    Y2 = YP

def state_HI_paramates():
    global AP, AV, A11, X1, X2, X3, X4, X5, X6,\
           DT, WV, UV, TTI, YYP
    AP = A2
    AV = 0
    A11 = 0
    X1 = 1
    X2 = 0
    X3 = 0
    X4 = 1
    X5 = 0
    X6 = 1
    DT = TT
    WV = 0
    UV = 0
    TTI = TI
    YYP = YP

def state_II_paramates():
```

```
    global AP, A11, UP, YP, TS, DT, TTT, WP, WV, AV, X1,\
      X2, X3, X4, X5, X6
    AP = 0
    A11 = 0
    UP = 0
    YP = 0
    TS = 0
    DT = TST / 5
    TTT = TR + TB + TI
    WP = 0
    WV = 0
    AV = 0
    X1 = 1
    X2 = 0
    X3 = 0
    X4 = 1
    X5 = 1
    X6 = 0

def state_II_paramates_2910():
    global WP, WV, AV, X1, X2, X3, X4, X5, X6
    WP = 0
    WV = 0
    AV = 0
    X1 = 1
    X2 = 0
    X3 = 0
    X4 = 1
    X5 = 1
    X6 = 0

def state_I_paramates():
    global AP, AV, A11, X1, X2, X3, X4, X5, X6, \
        WP, UP, YP, TS, DT, TTT
    AP = 0
    AV = A5 - A3
    A11 = 0
    X1 = 0
    X2 = 1
    X3 = 0
    X4 = 1
    X5 = 1
    X6 = 0
    WP = 0
    UP = 0
    YP = 0
    TS = 0
    DT = TST / 5
    TTT = TR + TB + TI

def state_AII_paramates():
    global AP, AV, A11, X1, X2, X3, X4, X5, X6,\
        UP, YP, DT, S
    AP = 0
    AV = A5 - A3
```

```
    A11 = 0
    X1 = 0
    X2 = 1
    X3 = 0
    X4 = 1
    X5 = 0
    X6 = 1
    UP = 0
    YP = 0
    DT = TT
    S = S2V

def state_AI_paramates():
    global AP, AV, A11, X1, X2, X3, X4, X5, X6,\
        DT, S, YV
    AP = A2
    AV = A5 - A3
    A11 = A1
    X1 = 0
    X2 = 1
    X3 = 0
    X4 = 1
    X5 = 0
    X6 = 1
    DT = TT
    S = SV
    YV = YV + DYV

def gen_list():
    UP_list.append(UP)
    UV_list.append(UV)
    P_list.append(P / MM)
    P1_list.append(P1 / MM)

    P2_list.append(P2 / MM)
    PL_list.append(PL / MM)
    QPV_list.append(-QPV * QQ)
    QE_list.append(QE * QQ)
    QE2_list.append(QE2 * QQ)
    return

def gen_riversion_list():
    UP_list[-1] = UP
    UV_list[-1] = UV
    P_list[-1] = P / MM
    P1_list[-1] = P1 / MM

    P2_list[-1] = P2 / MM
    PL_list[-1] = PL / MM
    QPV_list[-1] = -QPV * QQ
    QE_list[-1] = QE * QQ
    QE2_list[-1] = QE2 * QQ

def A_state():
    global NN, QI, PS, PLA, K, TT, DTV, PII, TAUP, TAUV,\
```

```
MU, EPSP, EPSV, RHO, ZT1, ZT2, ZT3, ZT4, ZT5, ZT6,\
C, D, H, L, L6, D1, H1, L1, D2, H2, L2, D3, H3, L3, D4, H4,\
L4, D5, H5, L5, MP, MV, TST, D0, L0, DH, LH, D7, D8, VLA,\
V20, Sc, Sic, SV, Z0, KL, KC, AA, LAMD, QD, PD, MM, QQ,\
Q, A0, A7, A8, M0, CL, VK0, V10, KE, K2K, KLM, KT, KTR,\
K10, B0P, B0V, B1, B2, B3, B4, B5, B6, B7, B8, B9, C7, C8,\
S1V, S2V, S3V, DTZ, R0P, R0V, R1, R2, R3, R4, R5, K1, K2,\
K3, K4, K5, K6, A1, A2, A3, A5, DT, VL, VK, P, P2, PL, P1,\
VK1, QE, U0, Y0, T0, QLV, DYV, DYP, TEE, FV, FSV, FLV,\
ET, EV, ES, EK, EL, EC, EO, UIM, J, HF, T_list, UP_list, \
UV_list, SP_list, P_list, P1_list, P2_list, PL_list, QA_list,\
QPV_list, QE_list, QE2_list, FP, FSP, FLP, WP, YP, UP,\
WV, YV, UV, P3, F0, FS0, W0, DY0, QPV, DVK, AP, AV, QL,\
QE2, A11, Q1, Q2, Q3, Q5, QT, ZZ, X1, X2, X3, X4, Q11, X5,\
X6, U12, DY1, U1, T, Y1, S, W1, U22, U2, DY2, Y2, W2, TB,\
SP, I, Z1, Z2, UVQ, EE, FF, GG, Q12, TVI, TV, SVV, N, TI, TTT,\
TR, KK, PP1, AS, TTI, YYP, TS, I1

print("State A")
state_A_paramates()
while True:
   DPr_4290()
   piston_moving_4020()
   backoil_moving_4140()
   computation_flow_4430()
   high_accumulator_4190()
   low_accumulator_4240()
   computation_EnergyLoss_4470()
   TR = TR + DT
   I1 += 1
   T_list.append(TR + TEE)
   SP_list.append(YP)
   gen_list()

   print('A:   {:<10.4f} {:<10.4f} {:<10.4f} {:<10.4f}'
  ' {:<10.4f} {:<10.4f}'.format(MM * VK, MM * VL, UP,
   Q * YP, UV, Q * YV))
   print('    {:<10.5f} {:<10.5f} {:<10.5f}'
   '{:<10.5f} {:<10.5f} {:<10.5f}'.format(P / MM, P1 / MM, P2 /
   MM, P3 / MM, WP, WV))
   print('    {:<10.4f} {:<10.5f} {:<10.5f}'
    '{:<10.5f} {:<10.5f}'.format(QQ * QI, -QQ * QPV, QQ *
   QE, QQ * QE2, -QQ * QL))
   if YP >= S:
      T = TR
      W1 = WP
      U1 = UP
      Y1 = YP
      computation_revisionI_4540()
      DPr_4290()
      TR = T
      UP = U1
      YP = S
      T_list[-1] = TR + TEE
      SP_list[-1] = YP
```

```
        gen_riversion_list()
        print('A:   {:<10.4f} {:<10.4f} {:<10.4f} {:<10.4f}'
        ' {:<10.4f} {:<10.4f}'.format(MM * VK, MM * VL, UP,
        Q * YP, UV, Q * YV))
        print('    {:<10.5f} {:<10.5f} {:<10.5f}'
        '{:<10.5f} {:<10.5f} {:<10.5f}'.format(P / MM, P1 / MM, P2 /
        MM, P3 / MM, WP, WV))
        break
    print("------------------------------")

def B_state():
    global NN, QI, PS, PLA, K, TT, DTV, PII, TAUP, TAUV,\
    MU, EPSP, EPSV, RHO, ZT1, ZT2, ZT3, ZT4, ZT5, ZT6, C,\
    D, H, L, L6, D1, H1, L1, D2, H2, L2, D3, H3, L3, D4, H4, L4,\
    D5, H5, L5, MP, MV, TST, D0, L0, DH, LH, D7, D8, VLA, V20,\
    Sc, Sic, SV, Z0, KL, KC, AA, LAMD, QD, PD, MM, QQ, Q, A0,\
    A7, A8, M0, CL, VK0, V10, KE, K2K, KLM, KT, KTR, K10, B0P,\
    B0V, B1, B2, B3, B4, B5, B6, B7, B8, B9, C7, C8, S1V, S2V, S3V,\
    DTZ, R0P, R0V, R1, R2, R3, R4, R5, K1, K2, K3, K4, K5, K6, A1, A2,\
    A3, A5, DT, VL, VK, P, P2, PL, P1, VK1, QE, U0, Y0, T0, QLV, DYV,\
    DYP, TEE, FV, FSV, FLV, ET, EV, ES, EK, EL, EC, EO, UIM, J, HF,
    T_list, UP_list, UV_list, SP_list, P_list, P1_list, P2_list, \
    PL_list, QA_list, QPV_list, QE_list, QE2_list, FP, FSP, FLP,\
    WP, YP, UP, WV, YV, UV, P3, F0, FS0, W0, DY0, QPV, DVK, AP, \
    AV, QL, QE2, A11, Q1, Q2, Q3, Q5, QT, ZZ, X1, X2, X3, X4, Q11,\
    X5, X6, U12, DY1, U1, T, Y1, S, W1, U22, U2, DY2,\
    Y2, W2, TB, SP, I, Z1, Z2, UVQ, EE, FF, GG, Q12, TVI, TV, SVV,\
    N, TI, TTT, TR, KK, PP1, AS, TTI, YYP, TS, I1, SR, URM

    print("State B")
    state_B_paramates()
    while True:
        DPr_4290()
        piston_moving_4020()
        valve_moving_4090()
        backoil_moving_4140()
        computation_flow_4430()
        high_accumulator_4190()
        low_accumulator_4240()
        computation_EnergyLoss_4470()
        if WP < 0: # Enter the return deceleration phase.
            TB += DT
            TV += DT
            I1 += 1
            T_list.append(TR + TB + TEE)
            SP_list.append(YP)
            gen_list()
        else:
            TR += DT
            TV += DT
            SR = YP
            URM = UP
            I1 += 1
            T_list.append(TR + TEE)
            SP_list.append(YP)
```

```
            gen_list()
        print('B:   {:<10.4f} {:<10.4f} {:<10.4f} {:<10.4f}'
        '{:<10.4f} {:<10.4f}'.format(MM * VK, MM * VL,
        UP, Q * YP, UV, Q * YV))
        print('    {:<10.5f} {:<10.5f} {:<10.5f}'
        '{:<10.5f} {:<10.5f} {:<10.5f}'.format(P / MM, P1 / MM,
        P2 / MM, P3 / MM, WP, WV))
        print('    {:<10.4f} {:<10.5f} {:<10.5f} {:<10.5f}'
        '{:<10.5f}'.format(QQ * QI, -QQ * QPV, QQ *
        QE, QQ * QE2, -QQ * QL))

        if YV >= S:
            T = TB
            W1 = WV
            U1 = UV
            Y1 = YV
            W2 = WP
            U2 = UP
            Y2 = YP
            computation_revisionII_4610()
            TB = T
            UV = U1
            YV = S
            UP = U2
            YP = Y2
            DPr_4290()
            T_list[-1] = TR + TB + TEE
            SP_list[-1] = YP
            print('B:   {:<10.4f} {:<10.4f} {:<10.4f} {:<10.4f}'
         '{:<10.4f} {:<10.4f}'.format(MM * VK, MM * VL,
         UP, Q * YP, UV, Q * YV))
            print('    {:<10.5f} {:<10.5f} {:<10.5f} {:<10.5f}'
         '{:<10.5f} {:<10.5f}'.format(P / MM, P1 / MM, P2 /
         MM, P3 / MM, WP, WV))
            print('    {:<10.4f} {:<10.5f} {:<10.5f}'
         '{:<10.5f} {:<10.5f}'.format(QQ * QI, -QQ * QPV, QQ
         * QE, QQ * QE2, -QQ * QL))
            print("Tr={:<8s}ms, Sr={:<8s}mm, "
             "Urm={:<8s}m/s".format(format_value(Q * TR),
             format_value(Q * SR), format_value(URM),
            ))
            break

def C_state():
    global NN, QI, PS, PLA, K, TT, DTV, PII, TAUP, TAUV, MU, EPSP,\
        EPSV, RHO, ZT1, ZT2, ZT3, ZT4, ZT5, ZT6, C, D, H, L, L6,\
        D1, H1, L1, D2, H2, L2, D3, H3, L3, D4, H4, L4, D5, H5,\
        L5, MP, MV, TST, D0, L0, DH, LH, D7, D8, VLA, V20, Sc, Sic,\
        SV, Z0, KL, KC, AA, LAMD, QD, PD, MM, QQ, Q, A0, A7, A8, \
        M0, CL, VK0, V10, KE, K2K, KLM, KT, KTR, K10, B0P, B0V, B1,\
        B2, B3, B4, B5, B6, B7, B8, B9, C7, C8, S1V, S2V, S3V, DTZ,\
        R0P, R0V, R1, R2, R3, R4, R5, K1, K2, K3, K4, K5, K6, A1, \
        A2, A3, A5, DT, VL, VK, P, P2, PL, P1, VK1, QE, U0, Y0, T0,\
        QLV, DYV, DYP, TEE, FV, FSV, FLV, ET, EV, ES, EK, EL, EC,\
        EO, UIM, J, HF, T_list, UP_list, UV_list, SP_list, P_list, \
```

```
        P1_list, P2_list, PL_list, QA_list, QPV_list, QE_list, \
        QE2_list, FP, FSP, FLP, WP, YP, UP, WV, YV, UV, P3, F0,\
        FS0, W0, DY0, QPV, DVK, AP, AV, QL, QE2, A11, Q1, Q2, Q3, \
        Q5, QT, ZZ, X1, X2, X3, X4, Q11, X5, X6, U12, DY1, U1, T, \
        Y1, S, W1, U22, U2, DY2, Y2, W2, TB, SP, I, Z1, Z2, UVQ, \
        EE, FF, GG, Q12, TVI, TV, SVV, N, TI, TTT, TR, KK, PP1, \
        AS, TTI, YYP, TS, I1, SR, URM, ELZ

    print("State C:")
    state_C_paramates()
    while True:
        DPi_4350()
        positive_mouth_interval_4750()
        ELZ = ELZ + P * Q11 * DT
        piston_moving_4020()
        valve_moving_4090()
        backoil_moving_4140()
        computation_flow_4430()
        high_accumulator_4190()
        low_accumulator_4240()
        TB += DT
        TV += DT
        I1 += 1
        T_list.append(TR + TB + TEE)
        SP_list.append(YP)
        gen_list()
        print('C:   {:<10.4f} {:<10.4f} {:<10.4f}'
        '{:<10.4f} {:<10.4f} {:<10.4f}'.format(MM * VK, MM *
        VL, UP, Q * YP, UV, Q * YV))
        print('    {:<10.5f} {:<10.5f} {:<10.5f}'
        '{:<10.5f} {:<10.5f} {:<10.5f}'.format(P / MM, P1 / MM, P2 /
         MM, P3 / MM, WP, WV))
        print('    {:<10.4f} {:<10.5f} {:<10.5f} {:<10.5f}'
        '{:<10.5f}'.format(QQ * QI, -QQ * QPV, QQ *
        QE, QQ * QE2, -QQ * QL))

        if Z1 >= Z0-DTZ:
            break

    print("------------------------------")

def D_state():
    global NN, QI, PS, PLA, K, TT, DTV, PII, TAUP, TAUV,\
    MU, EPSP, EPSV, RHO, ZT1, ZT2, ZT3, ZT4, ZT5, ZT6, C, D,\
    H, L, L6, D1, H1, L1, D2, H2, L2, D3, H3, L3, D4, H4, L4, D5,\
    H5, L5, MP, MV, TST, D0, L0, DH, LH, D7, D8, VLA, V20, Sc,\
    Sic, SV, Z0, KL, KC, AA, LAMD, QD, PD, MM, QQ, Q, A0, A7,\
    A8, M0, CL, VK0, V10, KE, K2K, KLM, KT, KTR, K10, B0P, B0V,\
    B1, B2, B3, B4, B5, B6, B7, B8, B9, C7, C8, S1V, S2V, S3V, DTZ,\
    R0P, R0V, R1, R2, R3, R4, R5, K1, K2, K3, K4, K5, K6, A1,\
    A2, A3, A5, DT, VL, VK, P, P2, PL, P1, VK1, QE, U0, Y0, T0, QLV,\
    DYV, DYP, TEE, FV, FSV, FLV, ET, EV, ES, EK, EL, EC, EO, UIM,\
    J, HF, T_list, UP_list, UV_list, SP_list, P_list, P1_list,\
   P2_list, PL_list, QA_list, QPV_list, QE_list, QE2_list,FP, FSP,\
    FLP, WP, YP, UP, WV, YV, UV, P3, F0, FS0, W0, DY0, QPV,\
```

```
DVK, AP, AV, QL, QE2, A11, Q1, Q2, Q3, Q5, QT, ZZ, X1, X2, X3,\
X4, Q11, X5, X6, U12, DY1, U1, T, Y1, S, W1, U22, U2, DY2, Y2, \
W2, TB, SP, I, Z1, Z2, UVQ, EE, FF, GG, Q12, TVI, TV, SVV, \
N, TI, TTT, TR, KK, YYP, TS, I1, SR, URM, ELZ, TVR

print("State D:")
state_D_paramates()
while True:
   DPi_4350()
   piston_moving_4020()
   valve_moving_4090()
   backoil_moving_4140()
   computation_flow_4430()
   high_accumulator_4190()
   low_accumulator_4240()
   computation_EnergyLoss_4470()
   TB += DT
   TV += DT
   I1 += 1
   T_list.append(TR + TB + TEE)
   SP_list.append(YP)
   gen_list()
   print('D:   {:<10.4f} {:<10.4f} {:<10.4f} {:<10.4f}'
   '{:<10.4f} {:<10.4f}'.format(MM * VK, MM * VL,
   UP, Q * YP, UV, Q * YV))
   print('    {:<10.5f} {:<10.5f} {:<10.5f} {:<10.5f}'
   '{:<10.5f} {:<10.5f}'.format(P / MM, P1 / MM, P2 /
   MM, P3 / MM, WP, WV))
   print('    {:<10.4f} {:<10.5f} {:<10.5f} '
   '{:<10.5f} {:<10.5f}'.format(QQ * QI, -QQ * QPV, QQ *
    QE, QQ * QE2, -QQ * QL))

   if YV >= S:
      T = TB
      W1 = WV
      U1 = UV
      Y1 = YV
      W2 = WP
      U2 = UP
      Y2 = YP
      computation_revisionII_4610()
      DPi_4350()
      TB = T
      UV = U1
      YV = S
      UP = U2
      YP = Y2
      T_list[-1] = TR + TB + TEE
      SP_list[-1] = YP
      gen_riversion_list()
      print('D:   {:<10.4f} {:<10.4f} {:<10.4f} {:<10.4f}'
      '{:<10.4f} {:<10.4f}'.format(MM * VK, MM * VL, UP,
       Q * YP,UV, Q * YV))
      print('    {:<10.5f} {:<10.5f} {:<10.5f} {:<10.5f}'
      '{:<10.5f} {:<10.5f}'.format(P / MM, P1 / MM, P2 /
```

```
        MM, P3 / MM, WP, WV))
        print("TR={:<8s} ms  Tb={:<8s} ms"
        .format(format_value(Q * TR),
        format_value(Q * TB)
        ))
        TVR = TV#Record the valve spool return reversal time.
        TV = 0
        break
    print("Tvr={:<8s} ms".format(format_value(TVR * Q)))
    print("------------------------------")

def E_state():
    global NN, QI, PS, PLA, K, TT, DTV, PII, TAUP, TAUV, MU, \
      EPSP, EPSV, RHO, ZT1, ZT2, ZT3, ZT4, ZT5, ZT6, C, D, \
      H, L, L6, D1, H1, L1, D2, H2, L2, D3, H3, L3, D4, H4,\
      L4, D5, H5, L5, MP, MV, TST, D0, L0, DH, LH, D7, D8,\
      VLA, V20, Sc, Sic, SV, Z0, KL, KC, AA, LAMD, QD, PD, \
      MM, QQ, Q, A0, A7, A8, M0, CL, VK0, V10, KE, K2K, KLM,\
      KT, KTR, K10, B0P, B0V, B1, B2, B3, B4, B5, B6, B7, \
      B8, B9, C7, C8, S1V, S2V, S3V, DTZ, R0P, R0V, R1, R2,\
      R3, R4, R5, K1, K2, K3, K4, K5, K6, A1, A2, A3, A5, DT, \
      VL, VK, P, P2, PL, P1, VK1, QE, U0, Y0, T0, QLV, DYV,\
      DYP, TEE, FV, FSV, FLV, ET, EV, ES, EK, EL, EC, EO, UIM,\
      J, HF, T_list, UP_list, UV_list, SP_list, P_list, P1_list, \
      P2_list, PL_list, QA_list, QPV_list, QE_list, QE2_list, \
      FP, FSP, FLP, WP, YP, UP, WV, YV, UV, P3, F0, FS0, W0, \
      DY0, QPV, DVK, AP, AV, QL, QE2, A11, Q1, Q2, Q3, Q5, QT,\
      ZZ, X1, X2, X3, X4, Q11, X5, X6, U12, DY1, U1, T, Y1, \
      S, W1, U22, U2, DY2, Y2, W2, TB, SP, I, Z1, Z2, UVQ, EE, \
      FF, GG, Q12, TVI, TV, SVV, N, TI, TTT, TR, KK, PP1, AS,\
      TTI, YYP, TS, I1, SR, URM, ELZ, SB

    print("State E:")
    state_E_paramates()
    while True:
      DPi_4350()
      piston_moving_4020()
      backoil_moving_4140()
      computation_flow_4430()
      high_accumulator_4190()
      low_accumulator_4240()
      computation_EnergyLoss_4470()
      TB = TB + DT
      I1 = I1 + 1
      T_list.append(TR + TB + TEE)
      SP_list.append(YP)
      gen_list()
      print('E:   {:<10.4f} {:<10.4f} {:<10.4f} {:<10.4f} '
      '{:<10.4f} {:<10.4f}'.format(MM * VK, MM *
      VL, UP, Q * YP, UV, Q * YV))
      print('     {:<10.5f} {:<10.5f} {:<10.5f} {:<10.5f}'
      '{:<10.5f} {:<10.5f}'.format(P / MM, P1 / MM, P2 /
       MM, P3 / MM, WP, WV))
      print('     {:<10.4f} {:<10.5f} {:<10.5f} {:<10.5f}'
     ' {:<10.5f}'.format(QQ * QI, -QQ * QPV, QQ * QE,
```

```
        QQ * QE2, -QQ * QL))

        if UP <= 0:
            computation_revisionIII_4690()
            DPi_4350()
            T_list[-1] = TR + TB + TEE
            SP_list[-1] = YP
            gen_riversion_list()
            print('E:   {:<10.4f} {:<10.4f} {:<10.4f}'
            '{:<10.4f} {:<10.4f} {:<10.4f}'.format(MM * VK,
            MM * VL, UP, Q * YP, UV, Q * YV))
            print('    {:<10.5f} {:<10.5f} {:<10.5f} {:<10.5f}'
            ' {:<10.5f} {:<10.5f}'.format(P / MM, P1 / MM, P2 /
            MM, P3 / MM, WP, WV))
            SB = SP - SR
            print('Tb: {:<8s} ms; Sp: {:<8s} mm; Sb: {:<8s} mm;'
            ' Sr: {:<8s} mm'.format(format_value(Q * TB),
                format_value(Q * SP),
                format_value(Q * SB),
                format_value(Q * SR)
            ))
            break
    print("------------------------------")

def F_state():
    global NN, QI, PS, PLA, K, TT, DTV, PII, TAUP, TAUV, MU, \
        EPSP, EPSV, RHO, ZT1, ZT2, ZT3, ZT4, ZT5, ZT6, C, D,\
        H, L, L6, D1, H1, L1, D2, H2, L2, D3, H3, L3, D4, H4, \
        L4, D5, H5, L5, MP, MV, TST, D0, L0, DH, LH, D7, D8,\
        VLA, V20, Sc, Sic, SV, Z0, KL, KC, AA, LAMD, QD, PD, \
        MM, QQ, Q, A0, A7, A8, M0, CL, VK0, V10, KE, K2K, KLM,\
        KT, KTR, K10, B0P, B0V, B1, B2, B3, B4, B5, B6, B7,\
        B8, B9, C7, C8, S1V, S2V, S3V, DTZ, R0P, R0V, R1, R2,\
        R3, R4, R5, K1, K2, K3, K4, K5, K6, A1, A2, A3, A5, DT, \
        VL, VK, P, P2, PL, P1, VK1, QE, U0, Y0, T0, QLV, DYV,\
        DYP, TEE, FV, FSV, FLV, ET, EV, ES, EK, EL, EC, EO, UIM, \
        J, HF, T_list, UP_list, UV_list, SP_list, P_list, P1_list,\
        P2_list, PL_list, QA_list, QPV_list, QE_list, QE2_list, \
        FP, FSP, FLP, WP, YP, UP, WV, YV, UV, P3, F0, FS0, W0,\
        DY0, QPV, DVK, AP, AV, QL, QE2, A11, Q1, Q2, Q3, Q5, QT,\
        ZZ, X1, X2, X3, X4, Q11, X5, X6, U12, DY1, U1, T, Y1, \
        S, W1, U22, U2, DY2, Y2, W2, TB, SP, I, Z1, Z2, UVQ, EE,\
        FF, GG, Q12, TVI, TV, SVV, N, TI, TTT, TR, KK, PP1, AS,\
        TTI, YYP, TS, I1, SR, URM, ELZ, SB, PMAX, VKMAX

    print("State F:")
    state_F_paramates()
    while True:
        DPi_4350()
        piston_moving_4020()
        backoil_moving_4140()
        computation_flow_4430()
        high_accumulator_4190()
        low_accumulator_4240()
        computation_EnergyLoss_4470()
```

```
    if N <= 0:
        if QE > 0:
            PMAX = P
            VKMAX = VK
            N = 1
    TI = TI + DT
    I1 += 1
    T_list.append(TTT + TI + TEE)
    SP_list.append(SP + YP)
    gen_list()
    print('F:   {:<10.4f} {:<10.4f} {:<10.4f} {:<10.4f}'
    '{:<10.4f} {:<10.4f}'.format(MM * VK, MM * VL,
    UP, Q * YP, UV, Q * YV))
    print('    {:<10.5f} {:<10.5f} {:<10.5f} {:<10.5f}'
    '{:<10.5f} {:<10.5f}'.format(P / MM, P1 / MM, P2 /
    MM, P3 / MM, WP, WV))
    print('    {:<10.4f} {:<10.5f} {:<10.5f}'
    '{:<10.5f} {:<10.5f}'.format(QQ * QI, -QQ * QPV, QQ *
    QE, QQ * QE2, -QQ * QL))
    if YP <= S:
        T = TI
        W1 = WP
        U1 = UP
        Y1 = YP
        computation_revisionI_4540()
        DPi_4350()
        TI = T
        UP = U1
        YP = S
        T_list[-1] = TTT + TI + TEE
        SP_list[-1] = SP + YP
        gen_riversion_list()
        print('F:   {:<10.4f} {:<10.4f} {:<10.4f} {:<10.4f}'
        '{:<10.4f} {:<10.4f}'.format(MM * VK, MM * VL,
        UP, Q * YP, UV, Q * YV))
        print('    {:<10.5f} {:<10.5f} {:<10.5f} {:<10.5f}'
        '{:<10.5f} {:<10.5f}'.format(P / MM, P1 / MM, P2 /
        MM, P3 / MM, WP, WV))
        print('Tr={:<8s} ms  Ti={:<8s} ms  Tb={:<8s}'
        'ms'.format(format_value(TR * Q),
        format_value(TI * Q), format_value(TB * Q)
        ))
        break
  print("------------------------------")

def H_state():
  global NN, QI, PS, PLA, K, TT, DTV, PII, TAUP, TAUV, MU, EPSP,\
  EPSV, RHO, ZT1, ZT2, ZT3, ZT4, ZT5, ZT6, C, D, H, L, L6, \
  D1, H1, L1, D2, H2, L2, D3, H3, L3, D4, H4, L4, D5, H5,\
  L5, MP, MV, TST, D0, L0, DH, LH, D7, D8, VLA, V20, Sc, Sic,\
  SV, Z0, KL, KC, AA, LAMD, QD, PD, MM, QQ, Q, A0, A7, A8,\
  M0, CL, VK0, V10, KE, K2K, KLM, KT, KTR, K10, B0P, B0V, \
  B1, B2, B3, B4, B5, B6, B7, B8, B9, C7, C8, S1V, S2V, S3V, \
  DTZ, R0P, R0V, R1, R2, R3, R4, R5, K1, K2, K3, K4, K5, K6,\
```

```
    A1, A2, A3, A5, DT, VL, VK, P, P2, PL, P1, VK1, QE, U0, Y0,\
    T0, QLV, DYV, DYP, TEE, FV, FSV, FLV, ET, EV, ES, EK, EL, \
    EC, EO, UIM, J, HF, T_list, UP_list, UV_list, SP_list ,P_list,\
    P1_list, P2_list, PL_list, QA_list, QPV_list, QE_list,\
    QE2_list, FP, FSP, FLP, WP, YP, UP, WV, YV, UV, P3, F0,\
    FS0, W0, DY0, QPV, DVK, AP, AV, QL, QE2, A11, Q1, Q2, Q3, Q5,\
    QT, ZZ, X1, X2, X3, X4, Q11, X5, X6, U12, DY1, U1, T, \
    Y1, S, W1, U22, U2, DY2, Y2, W2, TB, SP, I, Z1, Z2, UVQ,\
    EE, FF, GG, Q12, TVI, TV, SVV, N, TI, TTT, TR, KK, PP1, \
    AS, TTI, YYP, TS, I1, SR, URM, ELZ, SB, PMAX, VKMAX

    print("State H:")
    state_H_paramates()
    computation_revisionII_4610()
    DPi_4350()
    TI = T
    TVI = TV - DT
    UV = U1
    YV = S
    UP = U2
    YP = Y2
    S = -SP
    T_list[-1] = TTT + TI + TEE
    SP_list[-1] = SP + YP
    gen_riversion_list()
    print('H   {:<10.4f} {:<10.4f} {:<10.4f} {:<10.4f}'
    '{:<10.4f} {:<10.4f}'.format(MM * VK, MM *
    VL, UP, Q * YP, UV, Q * YV))
    print('    {:<10.5f} {:<10.5f} {:<10.5f} {:<10.5f}'
    '{:<10.5f} {:<10.5f}'.format(P / MM, P1 / MM, P2 /
    MM, P3 / MM, WP, WV))
    print("Valve Stop. Tvi={:<8s} ms"
    .format(format_value(TVI * Q)))

def G_state():
    global NN, QI, PS, PLA, K, TT, DTV, PII, TAUP, TAUV, MU, EPSP,\
        EPSV, RHO, ZT1, ZT2, ZT3, ZT4, ZT5, ZT6, C, D, H, L, L6, \
        D1, H1, L1, D2, H2, L2, D3, H3, L3, D4, H4, L4, D5, H5, L5,\
        MP, MV, TST, D0, L0, DH, LH, D7, D8, VLA, V20, Sc, Sic, SV,\
        Z0, KL, KC, AA, LAMD, QD, PD, MM, QQ, Q, A0, A7, A8, M0, CL,\
        VK0, V10, KE, K2K, KLM, KT, KTR, K10, B0P, B0V, B1, B2, B3,\
        B4, B5, B6, B7, B8, B9, C7, C8, S1V, S2V, S3V, DTZ, R0P, R0V, \
        R1, R2, R3, R4, R5, K1, K2, K3, K4, K5, K6, A1, A2, A3, A5,\
        DT, VL, VK, P, P2, PL, P1, VK1, QE, U0, Y0, T0, QLV, DYV, \
        DYP, TEE, FV, FSV, FLV, ET, EV, ES, EK, EL, EC, EO, UIM, J, \
        HF, T_list, UP_list, UV_list, SP_list, P_list, P1_list, \
        P2_list, PL_list, QA_list, QPV_list, QE_list, QE2_list, FP, \
        FSP, FLP, WP, YP, UP, WV, YV, UV, P3, F0, FS0, W0, DY0, QPV,\
        DVK, AP, AV, QL, QE2, A11, Q1, Q2, Q3, Q5, QT, ZZ, X1, X2,\
        X3, X4, Q11, X5, X6, U12, DY1, U1, T, Y1, S, W1, U22, U2, DY2,\
        Y2, W2, TB, SP, I, Z1, Z2, UVQ, EE, FF, GG, Q12, TVI, TV, SVV, \
        N, TI, TTT, TR, KK, PP1, AS, TTI, YYP, TS, I1, SR, URM, ELZ, \
        SB, PMAX, VKMAX, directtoI, PMIN

    print("State G:")
```

```
state_G_paramates()
while True:
    if N > 0:
        DPi_4350()
    else:
        DPr_4290()

    piston_moving_4020()
    valve_moving_4090()
    backoil_moving_4140()
    computation_flow_4430()
    high_accumulator_4190()
    low_accumulator_4240()
    computation_EnergyLoss_4470()
    TI = TI + DT
    TV = TV + KK * DT
    I1 = I1 + 1
    print("WP=", WP)
    T_list.append(TTT + TI + TEE)
    SP_list.append(SP + YP)
    gen_list()
    print('{}   {:<10.4f} {:<10.4f} {:<10.4f} {:<10.4f}'
    '{:<10.4f} {:<10.4f}'.format(AS, MM * VK, MM * VL,
    UP, Q * YP, UV, Q * YV))
    print('    {:<10.5f} {:<10.5f} {:<10.5f} {:<10.5f}'
    '{:<10.5f} {:<10.5f}'.format(P / MM, P1 / MM, P2 /
    MM, P3 / MM, WP, WV))
    print('    {:<10.4f} {:<10.5f} {:<10.5f}'
    '{:<10.5f} {:<10.5f}'.format(QQ * QI, -QQ * QPV,
    QQ * QE, QQ * QE2, -QQ * QL))

    if WP > 0:# Enter to H
        N = 0
        AP = A2
        X1 = 1
        X2 = 0
        AS = "H"
    if YV > SV: # State H ends
        H_state()
        directtoI = False
        break
    if YP <= S:#Impact completed
        T = TI
        W1 = WP
        U1 = UP
        Y1 = YP
        W2 = WV
        U2 = UV
        Y2 = YV
        computation_revisionII_4610()
        DPi_4350()
        if YV < S2V:#Impact completed,but the oil circuit
                # has not been switched. Enter to I
            P1 = P1 + AA * RHO * abs(UP)
            PP1 = 1
```

```
        TI = T
        UP = U1
        YP = S
        UV = U2
        YV = Y2
        UIM = -UP
        PMIN = P
        T_list[-1] = TTT + TI + TEE
        SP_list[-1] = SP + YP
        gen_riversion_list()
        print('{}   {:<10.4f} {:<10.4f} {:<10.4f}'
        '{:<10.4f} {:<10.4f} {:<10.4f}'.format(AS, MM *
        VK, MM * VL, UP, Q * YP, UV, Q * YV))
        print('    {:<10.5f} {:<10.5f} {:<10.5f} {:<10.5f}'
        '{:<10.5f} {:<10.5f}'.format(P / MM, P1 / MM, P2 /
        MM, P3 / MM, WP, WV))
        AV = 0
        KK = 0
        UV = 0
        DYV = 0
        FSV = 0
        FLV = 0
        directtoI = True
        break
    print("------------------------------")

def HI_state(): #state H'
    global NN, QI, PS, PLA, K, TT, DTV, PII, TAUP, TAUV, MU, EPSP, \
      EPSV, RHO, ZT1, ZT2, ZT3, ZT4, ZT5, ZT6, C, D, H, L, L6, \
      D1, H1, L1, D2, H2, L2, D3, H3, L3, D4, H4, L4, D5, H5, L5, \
      MP, MV, TST, D0, L0, DH, LH, D7, D8, VLA, V20, Sc, Sic, SV,\
      Z0, KL, KC, AA, LAMD, QD, PD, MM, QQ, Q, A0, A7, A8, M0, CL,\
      VK0, V10, KE, K2K, KLM, KT, KTR, K10, B0P, B0V, B1, B2, B3,\
      B4, B5, B6, B7, B8, B9, C7, C8, S1V, S2V, S3V, DTZ, R0P, R0V, \
      R1, R2, R3, R4, R5, K1, K2, K3, K4, K5, K6, A1, A2, A3, A5, \
      DT, VL, VK, P, P2, PL, P1, VK1, QE, U0, Y0, T0, QLV, DYV, DYP, \
      TEE, FV, FSV, FLV, ET, EV, ES, EK, EL, EC, EO, UIM, J, HF, \
      T_list, UP_list, UV_list, SP_list, P_list, P1_list, P2_list,\
      PL_list, QA_list, QPV_list, QE_list, QE2_list, FP, FSP, FLP,\
      WP, YP, UP, WV, YV, UV, P3, F0, FS0, W0, DY0, QPV, DVK, AP,\
      AV, QL, QE2, A11, Q1, Q2, Q3, Q5, QT, ZZ, X1, X2, X3, X4, Q11,\
      X5, X6, U12, DY1, U1, T, Y1, S, W1, U22, U2, DY2, Y2, W2, TB, \
      SP, I, Z1, Z2, UVQ, EE, FF, GG, Q12, TVI, TV, SVV, N, TI, TTT,\
      TR, KK, PP1, AS, TTI, YYP, TS, I1, SR, URM, ELZ, SB, PMAX, \
      VKMAX, directtoI, PMIN

    print("State H':")
    state_HI_paramates()
    while True:
      DPr_4290()
      piston_moving_4020()
      backoil_moving_4140()
      computation_flow_4430()
      high_accumulator_4190()
      low_accumulator_4240()
```

```
        computation_EnergyLoss_4470()
        TI = TI + DT
        I1 = I1 + 1
        T_list.append(TTT + TI + TEE)
        SP_list.append(SP + YP)
        gen_list()
        print("H'  {:<10.4f} {:<10.4f} {:<10.4f} {:<10.4f}"
        "{:<10.4f} {:<10.4f}".format(MM * VK, MM * VL, UP,
        Q * YP, UV, Q * YV))
        print('    {:<10.5f} {:<10.5f} {:<10.5f} {:<10.5f}'
        '{:<10.5f} {:<10.5f}'.format(P / MM, P1 / MM, P2 /
        MM, P3 / MM, WP, WV))
        print('    {:<10.4f} {:<10.5f} {:<10.5f} {:<10.5f}'
        '{:<10.5f}'.format(QQ * QI, -QQ * QPV, QQ * QE, QQ *
        QE2, -QQ * QL))
        if YP <= S:
          T = TI
          W1 = WP
          U1 = UP
          Y1 = YP
          computation_revisionI_4540()
          DPr_4290()
          TI = T
          UP = U1
          YP = S
          UIM = -UP
          PMIN = P
          YV = SV
          T_list[-1] = (TTT + TI + TEE)
          SP_list[-1] = (SP + YP)
          gen_riversion_list()
          print("H'  {:<10.4f} {:<10.4f} {:<10.4f}"
          "{:<10.4f} {:<10.4f} {:<10.4f}".format(MM * VK,
          MM * VL, UP, Q * YP, UV, Q * YV))
          print('    {:<10.5f} {:<10.5f} {:<10.5f}'
          '{:<10.5f} {:<10.5f} {:<10.5f}'.format(P / MM, P1 /
          MM, P2 / MM, P3 / MM, WP, WV))
          print("'Valve Lead', Time:{:<8s}ms, Stroke:{:<8s}mm"
          .format(format_value(Q * (TI - TTI)), format_value
           (Q * (YP - YYP))
          ))
          break
    print("TV={:<8s} ms".format(format_value(TV * Q)))
    print("-------------------------------")

def II_state():#State I'
    global NN, QI, PS, PLA, K, TT, DTV, PII, TAUP, TAUV, MU,\
      EPSP, EPSV, RHO, ZT1, ZT2, ZT3, ZT4, ZT5, ZT6, C, D,\
      H, L, L6, D1, H1, L1, D2, H2, L2, D3, H3, L3, D4, H4,\
      L4, D5, H5, L5, MP, MV, TST, D0, L0, DH, LH, D7, D8, \
      VLA, V20, Sc, Sic, SV, Z0, KL, KC, AA, LAMD, QD, PD, \
      MM, QQ, Q, A0, A7, A8, M0, CL, VK0, V10, KE, K2K, KLM,\
      KT, KTR, K10, B0P, B0V, B1, B2, B3, B4, B5, B6, B7, B8,\
      B9, C7, C8, S1V, S2V, S3V, DTZ, R0P, R0V, R1, R2, R3,\
```

```
        R4, R5, K1, K2, K3, K4, K5, K6, A1, A2, A3, A5, DT, VL, \
        VK, P, P2, PL, P1, VK1, QE, U0, Y0, T0, QLV, DYV, DYP, \
        TEE, FV, FSV, FLV, ET, EV, ES, EK, EL, EC, EO, UIM, J, \
        HF, T_list, UP_list, UV_list, SP_list, P_list, P1_list,\
        P2_list, PL_list, QA_list, QPV_list, QE_list, QE2_list, \
        FP, FSP, FLP, WP, YP, UP, WV, YV, UV, P3, F0, FS0, W0,\
        DY0, QPV, DVK, AP, AV, QL, QE2, A11, Q1, Q2, Q3, Q5, QT, \
        ZZ, X1, X2, X3, X4, Q11, X5, X6, U12, DY1, U1, T, Y1, \
        S, W1, U22, U2, DY2, Y2, W2, TB, SP, I, Z1, Z2, UVQ, EE,\
        FF, GG, Q12, TVI, TV, SVV, N, TI, TTT, TR, KK, PP1, AS, \
        TTI, YYP, TS, I1, SR, URM, ELZ, SB, PMAX, VKMAX, directtoI, \
        PMIN

    print("State I':")
    state_II_paramates()
    while True:
        DPr_4290()
        backoil_moving_4140()
        computation_flow_4430()
        high_accumulator_4190()
        low_accumulator_4240()
        ET = ET + P * QI * DT
        EL = EL + ((P1 - PL) * QL1 + (P2 - PL) * QL2 + (P - PL)
        * QLV) * DT
        TS = TS + DT#Calculate the time spent in pause.
        I1 = I1 + 1
        T_list.append(TTT + TS + TEE)
        SP_list.append(YP)
        gen_list()
        print("I'  {:<10.4f} {:<10.4f} {:<10.4f}"
        "{:<10.4f} {:<10.4f} {:<10.4f}".format(MM * VK, MM *
        VL, UP, Q * YP, UV, Q * YV))
        print('    {:<10.5f} {:<10.5f} {:<10.5f}'
        '{:<10.5f} {:<10.5f} {:<10.5f}'.format(P / MM, P1 /
        MM, P2 / MM, P3 / MM, WP, WV))
        print('    {:<10.4f} {:<10.5f} {:<10.5f}'
        '{:<10.5f} {:<10.5f}'.format(QQ * QI, -QQ *
        QPV, QQ * QE, QQ * QE2, -QQ * QL))

        if TS >= TST:#The pause ends.
            break
    TTT = TR + TB + TI + TST#Record the time of one cycle.
    print("TV={:<8s} ms".format(format_value(TV * Q)))
    print('TI={:<8s} ms  TR={:<8s} ms'.format(
        format_value(TI * Q),
        format_value(TR * Q)
    ))
    print("------------------------------")

def II_state_2910():#Transition from state I to state I'
    global NN, QI, PS, PLA, K, TT, DTV, PII, TAUP, TAUV, MU, \
        EPSP, EPSV, RHO, ZT1, ZT2, ZT3, ZT4, ZT5, ZT6, C, D, \
        H, L, L6, D1, H1, L1, D2, H2, L2, D3, H3, L3, D4, H4,\
        L4, D5, H5, L5, MP, MV, TST, D0, L0, DH, LH, D7, D8, \
        VLA, V20, Sc, Sic, SV, Z0, KL, KC, AA, LAMD, QD, PD, \
```

```
      MM, QQ, Q, A0, A7, A8, M0, CL, VK0, V10, KE, K2K, KLM,\
      KT, KTR, K10, B0P, B0V, B1, B2, B3, B4, B5, B6, B7, B8,\
      B9, C7, C8, S1V, S2V, S3V, DTZ, R0P, R0V, R1, R2, R3, \
      R4, R5, K1, K2, K3, K4, K5, K6, A1, A2, A3, A5, DT, VL,\
      VK, P, P2, PL, P1, VK1, QE, U0, Y0, T0, QLV, DYV, DYP, \
      TEE, FV, FSV, FLV, ET, EV, ES, EK, EL, EC, EO, UIM, J, \
      HF, T_list, UP_list, UV_list, SP_list, P_list, P1_list, \
      P2_list, PL_list, QA_list, QPV_list, QE_list, QE2_list, \
      FP, FSP, FLP, WP, YP, UP, WV, YV, UV, P3, F0, FS0, W0, \
      DY0, QPV, DVK, AP, AV, QL, QE2, A11, Q1, Q2, Q3, Q5, QT,\
      ZZ, X1, X2, X3, X4, Q11, X5, X6, U12, DY1, U1, T, Y1, S,\
      W1, U22, U2, DY2, Y2, W2, TB, SP, I, Z1, Z2, UVQ, EE, FF,\
      GG, Q12, TVI, TV, SVV, N, TI, TTT, TR, KK, PP1, AS, TTI,\
      YYP, TS, I1, SR, URM, ELZ, SB, PMAX, VKMAX, directtoI, \
      PMIN

    state_II_paramates_2910()
    while True:
      DPr_4290()
      backoil_moving_4140()
      computation_flow_4430()
      high_accumulator_4190()
      low_accumulator_4240()
      ET = ET + P * QI * DT
      EL = EL + ((P1 - PL) * QL1 + (P2 - PL) * QL2 + (P - PL)
      * QLV) * DT
      TS = TS + DT
      I1 = I1 + 1
      T_list.append(TTT + TS + TEE)
      SP_list.append(YP)
      gen_list()
      print("I'  {:<10.4f} {:<10.4f} {:<10.4f}"
      "{:<10.4f} {:<10.4f} {:<10.4f}".format(MM * VK, MM *
      VL, UP, Q * YP, UV, Q * YV))
      print('    {:<10.5f} {:<10.5f} {:<10.5f}'
      '{:<10.5f} {:<10.5f} {:<10.5f}'.format(P / MM, P1 /
      MM, P2 / MM, P3 / MM, WP, WV))
      print('    {:<10.4f} {:<10.5f} {:<10.5f} {:<10.5f}'
      '{:<10.5f}'.format(QQ * QI, -QQ * QPV, QQ *
      QE, QQ * QE2, -QQ * QL))

      if TS >= TST:
        break
    TTT = TR + TB + TI + TST
    print("TV={:<8s} ms".format(format_value(TV * Q)))
    print('TI={:<8s} ms  TR={:<8s} ms'.format(
      format_value(TI * Q),
      format_value(TR * Q)
    ))
    print("-------------------------------")

  def I_state():
    global NN, QI, PS, PLA, K, TT, DTV, PII, TAUP, TAUV, MU,\
      EPSP, EPSV, RHO, ZT1, ZT2, ZT3, ZT4, ZT5, ZT6, C, D,\
      H, L, L6, D1, H1, L1, D2, H2, L2, D3, H3, L3, D4, H4,\
```

```
    L4, D5, H5, L5, MP, MV, TST, D0, L0, DH, LH, D7, D8,\
    VLA, V20, Sc, Sic, SV, Z0, KL, KC, AA, LAMD, QD, PD,\
    MM, QQ, Q, A0, A7, A8, M0, CL, VK0, V10, KE, K2K, KLM,\
    KT, KTR, K10, B0P, B0V, B1, B2, B3, B4, B5, B6, B7, B8,\
    B9, C7, C8, S1V, S2V, S3V, DTZ, R0P, R0V, R1, R2, R3, \
    R4, R5, K1, K2, K3, K4, K5, K6, A1, A2, A3, A5, DT, VL,\
    VK, P, P2, PL, P1, VK1, QE, U0, Y0, T0, QLV, DYV, DYP,\
    TEE, FV, FSV, FLV, ET, EV, ES, EK, EL, EC, EO, UIM, J,\
    HF, T_list, UP_list, UV_list, SP_list, P_list, P1_list,\
    P2_list, PL_list, QA_list, QPV_list, QE_list, QE2_list, \
    FP, FSP, FLP, WP, YP, UP, WV, YV, UV, P3, F0, FS0, W0,\
    DY0, QPV, DVK, AP, AV, QL, QE2, A11, Q1, Q2, Q3, Q5, QT,\
    ZZ, X1, X2, X3, X4, Q11, X5, X6, U12, DY1, U1, T, Y1, \
    S, W1, U22, U2, DY2, Y2, W2, TB, SP, I, Z1, Z2, UVQ, EE, \
    FF, GG, Q12, TVI, TV, SVV, N, TI, TTT, TR, KK, PP1, AS, \
    TTI, YYP, TS, I1, SR, URM, ELZ, SB, PMAX, VKMAX, directtoI, \
    PMIN

print("State I:")
state_I_paramates()
while True:
    if YV > SV:
        TVI = TV #Record the impact reversal time
        II_state_2910() #Enter state I'
        break
    DPi_4350()
    valve_moving_4090()
    backoil_moving_4140()
    computation_flow_4430()
    high_accumulator_4190()
    low_accumulator_4240()
    ET = ET + P * QI * DT
    EV = EV + X2 * FV * DYV
    EK = EK + X2 * FLV * DYV
    EL = EL + ((P1 - PL) * QL1 + (P2 - PL) *
    QL2 + (P - PL) * QLV) * DT
    TS = TS + DT
    TV = TV + DT
    I1 = I1 + 1
    T_list.append(TTT + TS + TEE)
    SP_list.append(YP)
    gen_list()
    print('I   {:<10.4f} {:<10.4f} {:<10.4f}'
    '{:<10.4f} {:<10.4f} {:<10.4f}'.format(MM * VK, MM *
    VL, UP, Q * YP, UV, Q * YV))
    print('    {:<10.5f} {:<10.5f} {:<10.5f}'
    '{:<10.5f} {:<10.5f} {:<10.5f}'.format(P / MM, P1 /
    MM, P2 / MM, P3 / MM, WP, WV))
    print('    {:<10.4f} {:<10.5f}'
    '{:<10.5f} {:<10.5f} {:<10.5f}'.format(QQ * QI, -QQ *
    QPV, QQ * QE, QQ * QE2, -QQ * QL))

    if TS >= TST:
        break
TTT = TR + TB + TI + TST
```

```
    print('TI={:<8s} ms  TR={:<8s} ms'.format(
        format_value(TI * Q),
        format_value(TR * Q)
    ))
    print("------------------------------")

def AI_state():# State A'
    global NN, QI, PS, PLA, K, TT, DTV, PII, TAUP, TAUV, MU,\
        EPSP, EPSV, RHO, ZT1, ZT2, ZT3, ZT4, ZT5, ZT6, C, D,\
        H, L, L6, D1, H1, L1, D2, H2, L2, D3, H3, L3, D4, H4, \
        L4, D5, H5, L5, MP, MV, TST, D0, L0, DH, LH, D7, D8, \
        VLA, V20, Sc, Sic, SV, Z0, KL, KC, AA, LAMD, QD, PD, \
        MM, QQ, Q, A0, A7, A8, M0, CL, VK0, V10, KE, K2K, KLM,\
        KT, KTR, K10, B0P, B0V, B1, B2, B3, B4, B5, B6, B7,\
        B8, B9, C7, C8, S1V, S2V, S3V, DTZ, R0P, R0V, R1, R2,\
        R3, R4, R5, K1, K2, K3, K4, K5, K6, A1, A2, A3, A5, DT, \
        VL, VK, P, P2, PL, P1, VK1, QE, U0, Y0, T0, QLV, DYV, \
        DYP, TEE, FV, FSV, FLV, ET, EV, ES, EK, EL, EC, EO, UIM, \
        J, HF, T_list, UP_list, UV_list, SP_list, P_list, P1_list,\
        P2_list, PL_list, QA_list, QPV_list, QE_list, QE2_list, \
        FP, FSP, FLP, WP, YP, UP, WV, YV, UV, P3, F0, FS0, W0, DY0,\
        QPV, DVK, AP, AV, QL, QE2, A11, Q1, Q2, Q3, Q5, QT, ZZ, X1,\
        X2, X3, X4, Q11, X5, X6, U12, DY1, U1, T, Y1, S, W1, U22, \
        U2, DY2, Y2, W2, TB, SP, I, Z1, Z2, UVQ, EE, FF, GG, Q12, \
        TVI, TV, SVV, N, TI, TTT, TR, KK, PP1, AS, TTI, YYP, TS, \
        I1, SR, URM, ELZ, SB, PMAX, VKMAX, directtoI, PMIN, TVD, YVD

    print("state A'")
    state_AI_paramates()
    while True:
        DPr_4290()
        piston_moving_4020()
        valve_moving_4090()
        backoil_moving_4140()
        computation_flow_4430()
        high_accumulator_4190()
        low_accumulator_4240()
        TR = TR + DT
        TV = TV + DT
        I1 = I1 + 1
        T_list.append(TR + TEE)
        SP_list.append(YP)
        gen_list()
        print("A'  {:<10.4f} {:<10.4f} {:<10.4f} {:<10.4f}"
         "{:<10.4f} {:<10.4f}".format(MM * VK, MM * VL, UP,
         Q * YP, UV, Q * YV))
        print('    {:<10.5f} {:<10.5f} {:<10.5f} {:<10.5f}'
        '{:<10.5f} {:<10.5f}'.format(P / MM, P1 / MM, P2 /
        MM, P3 / MM, WP, WV))
        print('    {:<10.4f} {:<10.5f} {:<10.5f} {:<10.5f}'
        '{:<10.5f}'.format(QQ * QI, -QQ * QPV, QQ
        * QE, QQ * QE2, -QQ * QL))

        if YV >= S:
            T = TR
```

```
            W1 = WV
            U1 = UV
            Y1 = YV
            W2 = WP
            U2 = UP
            Y2 = YP
            computation_revisionII_4610()
            DPi_4350()
            TR = T
            TV = TV - DT
            UV = U1
            YV = S
            UP = U2
            YP = Y2
            TVI = TV
            T_list[-1] = TR + TEE
            SP_list[-1] = YP
            gen_riversion_list()
            print("A'  {:<10.4f} {:<10.4f} {:<10.4f}"
            "{:<10.4f} {:<10.4f} {:<10.4f}"
            .format(MM * VK, MM * VL, UP, Q * YP, UV, Q * YV))
            print('    {:<10.5f} {:<10.5f} {:<10.5f}'
            '{:<10.5f} {:<10.5f} {:<10.5f}'
            .format(P / MM, P1 / MM, P2 / MM, P3 / MM, WP, WV))
            print("Valve Delay (A' State): TvD={:<8s} ms,"
            "YvD={:<8s} mm ".format(format_value(Q * (TV - TVD)),
            format_value(Q * (YV - YVD))
            ))
            break

def AII_state(): #State A"
    global NN, QI, PS, PLA, K, TT, DTV, PII, TAUP, TAUV, MU,\
        EPSP, EPSV, RHO, ZT1, ZT2, ZT3, ZT4, ZT5, ZT6, C, D,\
        H, L, L6, D1, H1, L1, D2, H2, L2, D3, H3, L3, D4, H4, \
        L4, D5, H5, L5, MP, MV, TST, D0, L0, DH, LH, D7, D8, \
        VLA, V20, Sc, Sic, SV, Z0, KL, KC, AA, LAMD, QD, PD, \
        MM, QQ, Q, A0, A7, A8, M0, CL, VK0, V10, KE, K2K, KLM,\
        KT, KTR, K10, B0P, B0V, B1, B2, B3, B4, B5, B6, B7, B8,\
        B9, C7, C8, S1V, S2V, S3V, DTZ, R0P, R0V, R1, R2, R3,\
        R4, R5, K1, K2, K3, K4, K5, K6, A1, A2, A3, A5, DT, VL,\
        VK, P, P2, PL, P1, VK1, QE, U0, Y0, T0, QLV, DYV, DYP,\
        TEE, FV, FSV, FLV, ET, EV, ES, EK, EL, EC, EO, UIM, J,\
        HF, T_list, UP_list, UV_list, SP_list, P_list, P1_list,\
        P2_list, PL_list, QA_list, QPV_list, QE_list, QE2_list, \
        FP, FSP, FLP, WP, YP, UP, WV, YV, UV, P3, F0, FS0, W0,\
        DY0, QPV, DVK, AP, AV, QL, QE2, A11, Q1, Q2, Q3, Q5, QT,\
        ZZ, X1, X2, X3, X4, Q11, X5, X6, U12, DY1, U1, T, Y1, \
        S, W1, U22, U2, DY2, Y2, W2, TB, SP, I, Z1, Z2, UVQ, EE, \
        FF, GG, Q12, TVI, TV, SVV, N, TI, TTT, TR, KK, PP1, AS,\
        TTI, YYP, TS, I1, SR, URM, ELZ, SB, PMAX, VKMAX, directtoI, \
        PMIN, TVD, YVD

    print('A"')
    state_AII_paramates()
    while True:
```

```
DPi_4350()
valve_moving_4090()
backoil_moving_4140()
computation_flow_4430()
high_accumulator_4190()
low_accumulator_4240()
computation_EnergyLoss_4470()

TR = TR + DT
TV = TV + DT
I1 = I1 + 1
T_list.append(TR + TEE)
SP_list.append(YP)
gen_list()
print('A"  {:<10.4f} {:<10.4f} {:<10.4f} {:<10.4f}'
'{:<10.4f} {:<10.4f}'.format(MM * VK, MM * VL,
UP, Q * YP, UV, Q * YV))
print('    {:<10.5f} {:<10.5f} {:<10.5f} {:<10.5f}'
'{:<10.5f} {:<10.5f}'.format(P / MM, P1 / MM, P2
/ MM, P3 / MM, WP, WV))
print('    {:<10.4f} {:<10.5f} {:<10.5f} {:<10.5f} {:<10.5f}'
.format(QQ * QI, -QQ * QPV, QQ * QE, QQ * QE2, -QQ * QL))
print('Tr={:<8s} ms  Ti={:<8s} ms  Tb={:<8s} ms'
.format(format_value(TR * Q),
format_value(TI * Q),
  format_value(TB * Q)
))
if YV >= S2V: #The oil circuit has been switched.
  T = TV
  W1 = WV
  U1 = UV
  Y1 = YV
  computation_revisionI_4540()
  DPr_4290()
  if PP1 < 0:
    P1 = P1 + AA * RHO * abs(UIM)
  TV = T
  TR = TR - DT
  UV = U1
  YV = S
  TV -= DT
  T_list[-1] = TR + TEE
  SP_list[-1] = YP
  gen_riversion_list()
  print('A"  {:<10.4f} {:<10.4f} {:<10.4f} {:<10.4f}'
  '{:<10.4f} {:<10.4f}'.format(MM * VK, MM * VL,
  UP, Q * YP, UV, Q * YV))
  print('    {:<10.5f} {:<10.5f} {:<10.5f} {:<10.5f}'
  '{:<10.5f} {:<10.5f}'.format(P / MM, P1 / MM, P2 /
  MM, P3 / MM, WP, WV))
  print('"Valve Delay(A" State): TvD={:<8s} ms,'
  'YvD={:<8s} mm'.format(format_value(Q * (TV - TVD)),
  format_value(Q * (YV - YVD))
  ))
```

```
            print("------------------------------")
            break

    def format_value(value):
        if abs(value) >= 1e3 or 0 < abs(value) < 1e-3:
            return "{:8.3e}".format(value)
        else:
            return "{:8.3f}".format(value)

    def output_function():

        PT = 0
        P_list_end = P_list[-(I1+1):-1]
        T_list_end = T_list[-(I1+1):-1]

        for i in range(1, I1):
            PT = PT + P_list_end[i]* (T_list_end[i] - T_list_end[i - 1])

        PAVE = PT /(T_list_end[-1]-T_list_end[0])

        w = 10000
        print("Ei:{:<8s} J  f:{:<8s} Hz".format(format_value(EI)
        , format_value(F)))
        print("Eta:{:<8s}  EtaM:{:<8s}  EtaV:{:<8s}  EtaP:{:<8s}"
        .format(format_value(ETA), format_value(ETAM),
        format_value(ETAV), format_value(ETAP)))
        print("Pmax:{:<8s} MPa  Pmin:{:<8s} MPa  Pave:{:<8s} MPa"
        .format(format_value(PMAX / MM), format_value(PMIN / MM),
        format_value(PAVE / MM)))
        print("Urm:{:<8s} m/s  Uim:{:<8s} m/s"
        .format(format_value(URM), format_value(UIM)))
        print("Tr:{:<8s} ms  Tb:{:<8s} ms  Ti:{:<8s} ms"
        .format(format_value(TR * Q), format_value(TB * Q),
        format_value(TI * Q)))
        print("Tvr:{:<8s} ms  Tvi:{:<8s} ms"
        .format(format_value(TVR * Q), format_value(TVI * Q)))
        print("Sp:{:<8s} mm  Sr:{:<8s} mm  Sb:{:<8s} mm"
        .format(format_value(SP * Q), format_value(SR * Q),
        format_value(SB * Q)))
        print("Sc:{:<8s} mm  Sic:{:<8s} mm"
        .format(format_value(Sc * Q), format_value(Sic * Q)))
       print("Qi:{:<8s} L/min  Pa:{:<8s} MPa  PLa:{:<8s} MPa  K:{:<6s}"
        .format(format_value(QI * QQ), format_value(PS / MM),
        format_value(PLA / MM), format_value(K)))
       print("Mp:{:<8s} kg  Mv:{:<8s} kg  VK0:{:<8s} ml  Vla:{:<8s} ml"
        .format(format_value(MP), format_value(MV),
        format_value(MM * VK0), format_value(MM * VLA)))
        print("A1:{:<8s} cm^2  A2:{:<8s} cm^2"
        "A3:{:<8s} cm^2  A5:{:<8s} cm^2"
        .format(format_value(A1 * w), format_value(A2 * w),
        format_value(A3 * w), format_value(A5 * w)))
        print("TauP:{:<8s}  TauV:{:<8s}  EpsP:{:<8s}"
       "EspV:{:<8s}  MU:{:<8s}"
        .format(format_value(TAUP), format_value(TAUV),
        format_value(EPSP), format_value(EPSV),
```

```
    format_value(MU)))
    print("Zeta1:{:<8s} Zeta2:{:<8s} Zeta3:{:<8s}"
    "Zeta4:{:<8s} Zeta5:{:<8s} ""Zeta6:{:<8s}"
    .format(format_value(ZT1), format_value(ZT2),
    format_value(ZT3), format_value(ZT4), format_value(ZT5),
    format_value(ZT6)))
    print("Et:{:<8s}  Ev:{:<8s}  Es:{:<8s}  Ek:{:<8s}"
    .format(format_value(ET), format_value(EV), format_value(ES),
    format_value(EK)))
    print("Eo:{:<8s}  El:{:<8s}  Ec:{:<8s}"
    .format(format_value(EO), format_value(EL),
    format_value(EC)))

    return

if __name__ == '__main__':

    (NN, QI, PS, PLA, K, TT, DTV, PII, TAUP, TAUV, MU,
     EPSP, EPSV, RHO, ZT1, ZT2, ZT3, ZT4, ZT5, ZT6, C,
     D, H, L, L6, D1, H1, L1, D2, H2, L2, D3, H3, L3,
     D4, H4, L4, D5, H5, L5, MP, MV, TST, D0, L0, DH,
     LH, D7, D8, VLA, V20, Sc, Sic, SV, Z0, KL, KC, AA,
     LAMD, QD, PD, MM, QQ, Q, A0, A7, A8, M0, CL, VK0,
     V10, KE, K2K, KLM, KT, KTR, K10, B0P, B0V, B1, B2,
     B3, B4, B5, B6, B7, B8, B9, C7, C8, S1V, S2V, S3V,
     DTZ, R0P, R0V, R1, R2, R3, R4, R5, K1, K2, K3, K4,
     K5, K6, A1, A2, A3, A5, DT, VL, VK, P, P2, PL, P1,
     VK1, QE, U0, Y0, T0, QLV, DYV, DYP, TEE, FV, FSV,
     FLV, ET, EV, ES, EK, EL, EC, EO, UIM, J, HF, T_list,
     UP_list, UV_list, SP_list, P_list, P1_list, P2_list,
     PL_list, QA_list, QPV_list, QE_list, QE2_list, FP,
     FSP, FLP, WP, YP, UP, WV, YV, UV, P3, F0, FS0, W0,
     DY0, QPV, DVK, AP, AV, QL, QE2, A11, Q1, Q2, Q3, Q5,
     QT, ZZ, X1, X2, X3, X4, Q11, X5, X6, U12, DY1, U1,
     T, Y1, S, W1, U22, U2, DY2, Y2, W2, TB, SP, I, Z1,
     Z2, UVQ, EE, FF, GG, Q12, TVI, TV, SVV, N, TI, TTT,
     TR, KK, PP1, AS, TTI, YYP, TS, I1, SR, URM, ELZ, SB,
     PMAX, VKMAX, directtoI, PMIN, TVD, YVD, TVR) = initail()

    UP = K * UIM
    YP = 0
    TR = 0
    I1 = 0
    P1_list_stable2 = []
    P_list_stable2 = []
    QE_list_stable2 = []
    QPV_list_stable2 = []
    SP_list_stable2 = []
    UP_list_stable2 = []
    QT_list = []
    QE2_list_stable2 = []

    while True:
        A_state()
        B_state()
```

```
        C_state()
        D_state()
        E_state()
        F_state()
        G_state()
        if directtoI:
            I_state()
        else:
            HI_state()
            II_state()

        VK2 = VK
        J = J + 1
        print("N#", J, " VK1", format_value(MM * VK1), "  VK2",
        format_value(MM * VK2))
        F = 1 / TTT
        EI = 0.5 * MP * UIM ** 2
        print("F:{:<8s} Hz  Ei:{:<8s} J  Pmax:{:<8s} Mpa"
        " Pmin:{:<8s} Mpa  "
              " Sp:{:<8s} mm".format(
            format_value(F),
            format_value(EI),
            format_value(PMAX / MM),
            format_value(PMIN / MM),
            format_value(SP * Q)
        ))
        ETA = 0.5 * MP * UIM ** 2 / ET
        ETAM = 1 - (ES + EK + EO) / ET
        ETAV = (A1 - A2) * SP / QI / TTT
        ETAP = 1 - EC / ET
        print("Number******", J)
        print("Eta:{:<8s}  EtaM:{:<8s}  EtaV:{:<8s}  EtaP:{:<8s}"
            .format(format_value(ETA), format_value(ETAM),
                  format_value(ETAV), format_value(ETAP)))
        print("ETA3: {:<8s}".format(format_value(ETAM * ETAV
* ETAP)))

        print("Et:{:<8s}  Ev:{:<8s}  Es:{:<8s} Mpa  Ek:{:<8s}"
        .format(format_value(ET), format_value(EV),
        format_value(ES), format_value(EK)))
        print("Eo:{:<8s}  El:{:<8s}  Ec:{:<8s} Mpa"
        .format(format_value(EO), format_value(EL),
        format_value(EC)))
        DE = ET - EV - ES - EK - EO - EL - EC - ((1 - K) ** 2) * EI
        print("DE: {:<8s}".format(format_value(DE)))

        print("Tvr:{:<8s} ms  Tvi:{:<8s} ms".format(
            format_value(TVR * Q), format_value(TVI * Q)))

        if abs(VK2 - VK1) <= DTV:
            output_function()
            draw_picture()
            break
        else:
            VK1 = VK2
```

```
        ET = 0
        EV = 0
        ES = 0
        EK = 0
        EO = 0
        EL = 0
        EC = 0
        if YV >= SV:
            TVI = TV
            print("A")
            UP = K * UIM
            YP = 0
            TR = 0
            I1 = 0
            TEE += TTT
            continue
        else:
            TR = 0
            YP = 0
            YVD = YV
            TVD = TV
            print("TVD = ", TV)
            I1 = 0
            TEE += TTT
            if YV >= S2V:
                UP = K * UIM
            else:
                AII_state()
            AI_state()
```

2. Code of Paramaters2.py

```
import numpy as np
import pandas as pd

def read_excel_parameters(excel_file):
    df = pd.read_excel(excel_file, header=None)

    parameters = {}

    for index, row in df.iterrows():
        parameter_name = row[0]
        parameter_value = row[1]
        parameter_value = float(parameter_value)
        parameters[parameter_name] = parameter_value

    return parameters

def parameters60(parameters_dict):
    QI = parameters_dict['QI']
    PS = parameters_dict['PS']
    PLA = parameters_dict['PLA']
    K = parameters_dict['K']
    TT = parameters_dict['TT']
    DTV = parameters_dict['DTV']
```

```
    return QI, PS, PLA, K, TT, DTV

def parameters80(parameters_dict):
    PII = 3.1416
    TAUP = parameters_dict['TAUP']
    TAUV = parameters_dict['TAUV']
    MU = parameters_dict['MU']
    EPSP = parameters_dict['EPSP']
    EPSV = parameters_dict['EPSV']
    RHO = parameters_dict['RHO']
    ZT1 = parameters_dict['ZT1']
    ZT2 = parameters_dict['ZT2']
    ZT3 = parameters_dict['ZT3']
    ZT4 = parameters_dict['ZT4']
    ZT5 = parameters_dict['ZT5']
    ZT6 = parameters_dict['ZT6']
    C = parameters_dict['C']
    return PII, TAUP, TAUV, MU, EPSP, EPSV, RHO,\
    ZT1, ZT2, ZT3, ZT4, ZT5, ZT6, C

def parameters100(parameters_dict):
    D = parameters_dict['D']
    H = parameters_dict['H']
    L = parameters_dict['L']
    L6 = parameters_dict['L6']
    D1 = parameters_dict['D1']
    H1 = parameters_dict['H1']
    L1 = parameters_dict['L1']
    D2 = parameters_dict['D2']
    H2 = parameters_dict['H2']
    L2 = parameters_dict['L2']
    return D, H, L, L6, D1, H1, L1, D2, H2, L2

def parameters120(parameters_dict):
    D3 = parameters_dict['D3']
    H3 = parameters_dict['H3']
    L3 = parameters_dict['L3']
    D4 = parameters_dict['D4']
    H4 = parameters_dict['H4']
    L4 = parameters_dict['L4']
    D5 = parameters_dict['D5']
    H5 = parameters_dict['H5']
    L5 = parameters_dict['L5']
    return D3, H3, L3, D4, H4, L4, D5, H5, L5

def parameters140(parameters_dict):
    MP = parameters_dict['MP']
    MV = parameters_dict['MV']
    TST = parameters_dict['TST']
    return MP, MV, TST

def parameters160(parameters_dict):
    D0 = parameters_dict['D0']
    L0 = parameters_dict['L0']
    DH = parameters_dict['DH']
```

```
    LH = parameters_dict['LH']
    D7 = parameters_dict['D7']
    D8 = parameters_dict['D8']
    return D0, L0, DH, LH, D7, D8

def parameters180(parameters_dict):
    VLA = parameters_dict['VLA']
    V20 = parameters_dict['V20']
    return VLA, V20

def parameters200(parameters_dict):
    Sc = parameters_dict['Sc']
    Sic = parameters_dict['Sic']
    SV = parameters_dict['SV']
    Z0 = parameters_dict['Z0']
    return Sc, Sic, SV, Z0

def parameters220(parameters_dict):
    KL = parameters_dict['KL']
    KC = parameters_dict['KC']
    AA = parameters_dict['AA']
    return KL, KC, AA

def parameters240(parameters_dict):
    LAMD = parameters_dict['LAMD']
    QD = parameters_dict['QD']
    PD = parameters_dict['PD']
    return LAMD, QD, PD

def parameters260():
    MM = 1000000
    QQ = 60000
    Q = 1000
    return MM, QQ, Q

def initialize():
    QE = 0
    U0 = 0
    Y0 = 0
    T0 = 0
    QLV = 0
    DYV = 0
    DYP = 0
    TEE = 0
    FV = 0
    FSV = 0
    FLV = 0
    ET = 0
    EV = 0
    ES = 0
    EK = 0
    EL = 0
    EC = 0
    EO = 0
    UIM = 0
```

```
    J = 0
    HF = 0
    T_list = []
    UP_list = []
    UV_list = []
    SP_list = []
    P_list = []
    P1_list = []
    P2_list = []
    PL_list = []
    QA_list = []
    QPV_list = []
    QE_list = []
    QE2_list = []

    return (QE, U0, Y0, T0, QLV, DYV, DYP, TEE, FV, FSV, FLV, ET,
         EV, ES, EK, EL, EC, EO, UIM, J, HF, T_list, UP_list,
         UV_list, SP_list, P_list, P1_list, P2_list, PL_list,
         QA_list, QPV_list, QE_list, QE2_list)

def coefficient(QI, PS, PLA, TT, DTV, PII, TAUP, TAUV, MU, EPSP,
EPSV, RHO,
     ZT1, ZT2, ZT3, ZT4, ZT5, ZT6, C, D, H, L, L6, D1, H1,
     L1, D2, H2, L2, D3, H3, L3, D4, H4, L4, D5, H5, L5, D0,
     L0, DH, LH, D7, D8, VLA, V20, SV, Z0, KL, KC, LAMD,
     QD, PD, MM, QQ):

    QI = QI / QQ
    PS = MM * PS
    PLA = MM * PLA
    VLA = VLA / MM
    DTV = DTV / MM
    V20 = V20 / MM
    A0 = (PII * D0 ** 2) / 4
    A7 = (PII * D7 ** 2) / 4
    A8 = (PII * D8 ** 2) / 4
    M0 = RHO * A0 * L0
    CL = PLA * VLA ** 1.4
    V10 = PII * DH ** 2 * LH / 4
    VK0 = V10 + V20
    VK = VK0
    KE = VK0 * KC * KL / (V10 * KL + VK0 * KC)
    K2K = V20 * KC / (V10 * KC + V10 * KL + V20 * KC)
    KLM = LAMD * RHO * L0 / D0 / 2
    KT = QD / QQ / (MM * PD)**0.5
    KTR = A0 * A0 / KT / KT
    K10 = A0 * (KLM + KTR)
    B0P = PII * MU / (1 - EPSP ** 2) ** 0.5
    B0V = PII * MU / (1 - EPSV ** 2) ** 0.5
    B1 = B0P * (D * L / H + D * L6 / H + D1 * L1 / H1 + D2 * L2 / H2)
    B2 = PII * (D2 * H2 - D * H) / 2
    B3 = PII * (D * H - D1 * H1) / 2
    B4 = TAUP * (D * L + D2 * L2)
    B5 = TAUP * (D * L6 + D1 * L1)
    B6 = B0V * (D3 * L3 / H3 + 2 * D4 * L4 / H4 + D5 * L5 / H5)
```

```
B7 = PII * D3 * H3 / 2
B8 = TAUV * (D3 * L3 + 2 * D4 * L4)
B9 = TAUV * D5 * L5
C7 = C * PII * D4 * (2 / RHO) ** 0.5
C8 = 1 / C7 ** 2
S1V = (SV - Z0) / 2
S2V = S1V + Z0
S3V = SV / 2
DTZ = Z0 / 9
R0P = PII * (1 + 1.5 * EPSP ** 2) / 12 / MU
R0V = PII * (1 + 1.5 * EPSV ** 2)
R1 = R0P * (D * H ** 3 / L6 + D1 * H1 ** 3 / L1)
R2 = PII * (D1 * H1 - D * H) / 2
R3 = R0P * (D * H ** 3 / L + D2 * H2 ** 3 / L2)
R4 = PII * (D * H - D2 * H2) / 2
R5 = R0V * (D3 * H3 ** 3 / L3 + 2 * D4 * H4 ** 3 / L4 +
 D5 * H5 ** 3 / L5)
K1 = ZT1 * RHO / A8 ** 2 / 2
K2 = ZT2 * RHO / A8 ** 2 / 2
K3 = ZT3 * RHO / A8 ** 2 / 2
K4 = ZT4 * RHO / A8 ** 2 / 2
K5 = ZT5 * RHO / A7 ** 2 / 2
K6 = ZT6 * RHO / A8 ** 2 / 2
A1 = PII * ((D - 2 * H) ** 2 - (D1 - 2 * H1) ** 2) / 4
A2 = PII * ((D - 2 * H) ** 2 - (D2 - 2 * H2) ** 2) / 4
A3 = PII * (D3 - 2 * H3) ** 2 / 4
A5 = PII * (D5 - 2 * H5) ** 2 / 4
DT = TT
VL = VLA
VK = VK0 * (1 + PS / KE)
P = PS
P2 = P
PL = PLA
P1 = PL
VK1 = VK
return (QI, PS, PLA, VLA, DTV, V20, A0, A7, A8, M0, CL,
     VK0, V10, KE, K2K, KLM, KT, KTR, K10, B0P,
     B0V, B1, B2, B3, B4, B5, B6, B7, B8, B9, C7, C8, S1V,
     S2V, S3V, DTZ, R0P, R0V, R1, R2, R3, R4, R5, K1, K2,
     K3, K4, K5, K6, A1, A2, A3, A5, DT, VL, VK, P, P2, PL,
     P1, VK1)
```

GPSR Compliance

The European Union's (EU) General Product Safety Regulation (GPSR) is a set of rules that requires consumer products to be safe and our obligations to ensure this.

If you have any concerns about our products, you can contact us on ProductSafety@springernature.com

In case Publisher is established outside the EU, the EU authorized representative is:

Springer Nature Customer Service Center GmbH
Europaplatz 3
69115 Heidelberg, Germany

Batch number: 10725802

Printed by Printforce, the Netherlands